BestMasters

Jessica Andel

Sense(s) of Heimat

Plurilocal Self-Location and
Emotional Geographies through the
Lens of International Migration

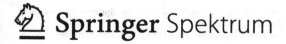

Springer Spektrum

Jessica Andel
Nieder-Olm, Germany

Die vorliegende Arbeit wurde im Jahr 2021 als Masterabschlussarbeit im Rahmen des Master of Arts Studiengangs „Human Geography: Globalisation, Media, and Culture" am Geographischen Institut der Johannes Gutenberg-Universität Mainz von Dr. phil. Elisabeth Sommerlad (Erstgutachterin) und Univ. Prof. Dr. Veronika Cummings (Zweitgutachterin) angenommen.

ISSN 2625-3577   ISSN 2625-3615 (electronic)
BestMasters
ISBN 978-3-658-38984-0   ISBN 978-3-658-38985-7 (eBook)
https://doi.org/10.1007/978-3-658-38985-7

Responsible Editor: Marija Kojic
This Springer Spektrum imprint is published by the registered company Springer Fachmedien Wiesbaden GmbH, part of Springer Nature.
The registered company address is: Abraham-Lincoln-Str. 46, 65189 Wiesbaden, Germany

Contents

Challenging a Static Conception of *Heimat*

1

"For the most part one goes much more with times/ by going against the times/ in recent times it has/ become common practice/ to go against the times/ so that in the end the going-against-the-times/ has again become a going-with-the-times/ that is why recently some are going with the times/ in the original sense of the idea/ only to actually go against the times/ in their very own way/ and thereby above all/ in the end to more easily go with the times again" (JONKE, trans. Vazulik, 1994 [1969], pp. 66 f.).

The notion *'Heimat'* is currently omnipresent in German speaking areas (c.f. BÖNISCH et al., 2020b, p. 1; EGGER, 2020, p. 23; WEBER et al., 2019, p. 5). It is a concept that is shared within German-speaking contexts, despite some local, historical, and social particularities (BLICKLE, 2002, p. 1). When German speakers talk about *Heimat*, they often refer to a certain quality of relationship, which is characterized by strong emotional attachment to a specific place such as the place of origin or residence, but also to larger spatial units such as a region or a nation-state, which are associated with a social community, particular commonalities, familiarity, habits and rules (MITZSCHERLICH, 2019, p. 183). The historian Peter BLICKLE states that *Heimat* is a crucial aspect in the self-perception of German-language cultures (2002, p. 1). As a seemingly unique and specific German phenomenon, *Heimat* is often considered as untranslatable into other languages and contexts (BLICKLE, 2002, p. 2; c.f. KÖSTLIN, 2010, p. 21).[1]

---

[1] There is no English or French equivalent for *Heimat*. Due to centuries of influence of German on Slavic languages there are, however, *Heimat* equivalents such as the Slovenian, Croatian, and Serbian word *dòmovina* and the Czech word *domov*. The qualities of the Russian word *rodina* are also close to the German concept of *Heimat*, although there are differences (BLICKLE, 2002, p. 2). However, it remains open whether or not there are concepts in other European or non-European languages with similar or equal semantic contents. A more detailed exploration hereto can be found in the article of BRUNS and MÜNDERLEIN (2019) on international concepts on human-place-relations.

© The Author(s), under exclusive license to Springer Fachmedien Wiesbaden GmbH, part of Springer Nature 2022
J. Andel, *Sense(s) of Heimat*, BestMasters,
https://doi.org/10.1007/978-3-658-38985-7_1

"If we consider the various translations of Heimat into English, we find such diverse results as "home," "homeland," "fatherland," "nation," "nation-state," "hometown," "paradise," "Germany," "Austria," "Switzerland," "Liechtenstein," "native region," "native landscape," "native soil," "birthplace," and "homestead"—and the list could be continued" (BLICKLE, 2002, p. 4).

From this multiplicity and contextuality of translations *Heimat* reveals its ambiguous quality. Even in German contexts it is problematic to define *Heimat* as this term does not entail universal meanings, but rather it is highly charged by subjectivity and irrationality (c.f. BASTIAN, 1995, pp. 23 f.; BLICKLE, 2002, pp. 12 f.). It is used in a variety of contexts, whether on the packaging of regional food products, in magazines on the idyllic life in the countryside, cookbooks with a focus on traditional dishes, songs about one's hometown, or films and TV productions about dramatic stories which take place in a beautiful mountain landscape—and this list goes on and on. *Heimat* is commonly associated with regionality, rural life, nature, ecological sustainability, and local traditions. Thereby, the notion is charged by positive, emotional, and aesthetic qualities (c.f. EGGER, 2020, p. 24; c.f. HÄNEL, 2020, p. 69; WEBER et al., 2019, p. 5). However, it can be observed that *Heimat* appears wave-like in the public discourse (WEBER et al., 2019, p. 7). BLICKLE notes that "[i]nvocations of Heimat [...] always turn up where deep socioeconomic, ontological, psychological, and political shifts, fissures, and insecurities occur. Heimat buries areas of repressed anxiety" (2002, pp. 13 f.). Especially in times of globalization and different situations of (global) crisis, references to and a longing for *Heimat* increase. This may be because *Heimat* represents a retreat from complexity, uncertainty, and alienation, as it seems to offer orientation, continuity, security, familiarity, community, and identity (c.f. BÖNISCH et al., 2020b, pp. 8 f.; SEIFERT, 2010b, p. 20). Thus, *Heimat* can be considered a fixed reference point in a globalized world (WEBER et al., 2019, p. 5). This orientation towards fixed places with the prospect of simplification and satisfaction matches the current "age of nostalgia", postulated by sociologist and philosopher Zygmunt BAUMAN in his work *Retrotopia* (2017). The search for orientation is often premised on a romanticized past and is thus directed towards retrograde practices and thinking patterns (BÖNISCH et al., 2020b, p. 9; c.f. BAUMAN, 2017, p. 5). In the words of KÖSTLIN, *Heimat* functions as an alternative program in order to cope with modernity. Moreover, he believes that modernity can only be understood through the contrast of traditional ideas of *Heimat* (2010, p. 36; c.f. opening quote by JONKE, 1994 [1969]).

Heimat is neither an objective nor a natural concept. It is rather constructed through interpretation along emotional, sociopolitical, and ideological loadings

(SEIFERT, 2010b, p. 20). This ambiguousness and the concept's inherent multiplicity make *Heimat* vulnerable to appropriation and instrumentalization by diverse (political) forces (c.f. BÖNISCH et al., 2020b, pp. 1 f.). The contested character of the concept is reflected, for instance, by vivid public discussions on key questions such as: What is *Heimat*? And who belongs to it? Future-oriented perspectives aspire, for example, to facilitate active participation in local matters and sociopolitical processes. In this sense *Heimat* is understood as inclusive and can be actively created. By contrast, *Heimat* also contains exclusionary potentials (EGGER, 2020, p. 25). Especially after increased numbers of refugee arrivals in 2015, the notion of *Heimat* is regularly (re)produced in political discourses. In particular, conservative as well as right-wing populist parties and movements use *Heimat* as a political buzzword with regard to emotionally charged topics such as migration and integration of refugees. In this context, *Heimat* appears to be a static concept, which is exposed to the risk of dramatic transformations and loss due to migration and associated negative changes such as processes of defamiliarization (WEBER et al., 2019, p. 5; c.f. YILDIZ and HILL, 2014b, p. 10). Thus, the German concept of *Heimat* appears to be hardly compatible with migration, as it "[...] carries a rich set of cultural and ideological connotations that combine notions of belonging and identity with affective attachment to a specific place or region" (EIGLER, 2012, p. 27). The aim of this thesis is to demonstrate that migration and a sense of *Heimat* do not antagonize one another—static conceptions of *Heimat* are hereby to be challenged.

The perception of *Heimat* as the antithesis to processes of globalization, mobility, and migration, is not a new phenomenon as the history of the concept shows (c.f. MITCHELL, 2000, pp. 264 f.). In order to understand the controversial discussions of *Heimat* as well as to critically scrutinize the concept in its contemporary meanings and dimensions, it is worth beginning with the concept's origin, usages and shifts in the course of German history, as "historical impulses have by no means completely disappeared yet, [...] but rather still continue to have an effect on present times" (BAUSINGER, trans., 1986, p. 107).[2]

---

[2] Most of the literature on *Heimat* is written in German. English translations of German quotations, if not otherwise noted, are done by the author (marked with 'trans.'). Translations of German terminologies and book titles are also done by the author with respect to common and contextual translations.

1.1   The German Idea of *Heimat*

The term can be traced back to the Old High German word '*heimôte/heimôti*' (WEBER et al., 2019, p. 6). For a long time the notion was used as a religious metaphor of the longed-for paradise in the beyond [*himmlische Heimat*], to which every devout Christian, irrespective of their place of origin, can refer (BAUSINGER, 1986, pp. 91 f.). In the 12ᵗʰ century, the connotation of the term shifted to secular references, such as harsh living conditions as well as a limited but familiar environment (WEBER, et al., 2019, p. 6). Early on, however, the semantic content of *Heimat* is not homogenous. This is highlighted by the fact, that "even though Heimat always constitutes the *antonym to foreign places* [*die Fremde*], the *spatial extension* of Heimat ranges from the entire country [*Land*], over the region [*Landstrich*] and locality [*Ort(schaft)*], to the home [*Haus/Wohnung*]" (BAUSINGER, trans., 1986, p. 91). *Heimat* is not defined by definite boundaries in an outward direction, but it is rather defined by what is not *Heimat* (c.f. JOISTEN, 2003, p. 19). However, for centuries the concept as a neuter noun [*das Heimat*] was specifically linked in a legal sense to property in the form of house and farm. To be in possession of *Heimat* was connected to legal rights (e.g., right of residence) and entitlements to benefit (e.g., support provided by the municipality in situations of poverty) but also to obligations (e.g., economic activity), which were regulated under the so-called *Heimatrecht* [*Heimat* right] in German-speaking areas. The *Heimatrecht*, which originated in the 16ᵗʰ century, was based on the idea of a patriarchal and static society in which *Heimat* could be acquired, for instance, by inheritance and marriage. Homeless people, servants, day laborers and other people without possessions were considered as *heimatlos* [*heimat*less], and thus were excluded from these rights but also from associated obligations. Those who were not able to fulfill the obligations or those who emigrated into *die Fremde* risked losing their *Heimatrecht* (BAUSINGER, 1986, pp. 92 f.; c.f. BASTIAN, 1995, pp. 101 ff.; KÖSTLIN, 2010, p. 32; BÖNISCH et al., 2020b, pp. 2 f.). Thus, *Heimat* was a polarized and exclusive concept that defined the inside/insiders and outside/outsiders by the status of possession (c.f. BLICKLE, 2002, p. 12). BAUSINGER concludes that *Heimat* then and now is contingent on social and sociopolitical framework conditions (1986, p. 94).

From idyllic landscape to national identity
With industrialization in the second half of the 19ᵗʰ century and the associated fundamental societal and economic upheavals, such as the shift from a focus on agriculture to an industrial society, the decline of rural life due to urbanization and the increase of geographic and social mobility (BASTIAN, 1995, p. 122), the

concept of the *Heimatrecht* became unsuitable (BAUSINGER, 1986, p. 94). *Heimat* as a synonym for paternal inheritance (land property) remains in use for a long time especially in Southern German dialects as well as in Austria and Switzerland (BASTIAN, 1995, p. 104). The concept of *Heimat*, however, shifted towards sentimental and pathetic connotations (WEBER et al., 2019, p. 6), "[...] particularly in the period after 1871 when the German Empire prospered politically, economically, socially, and culturally" (HÄGELE, 2021, p. 132). Important insights into the emotional meaning and understanding of *Heimat* can be found in written sources of German emigrants, who were in search of better income opportunities in Australia, North and South America. They experienced existential challenges as well as difficulties to adapt to new and different cultural environments. Letters of these emigrants, which were addressed to those who stayed in Europe, reveal a longing for their *Heimat* [*Heimweh*][3] and their fears to fall into oblivion. This indicates firstly that *Heimat* is not only characterized by possession and locality but, above all, by commonalities and social aspects, such as family, kindred, and friends (BASTIAN, 1995, pp. 82 f.), and secondly that the importance of *Heimat* increases when reminisced or lost (c.f. SZEJNMANN, 2012, pp. 122 f.). A fundamental semantic change of the concept of *Heimat* evolved, however, from the positive and utopian image of *Heimat* of romanticists and the intellectual bourgeoisie. In rejecting reaction to the effects of the industrialization on their privileges and traditional structures, the urban middle class associated *Heimat* with rural folk culture, idyllic landscapes, and pristine nature (BASTIAN, 1995, pp. 121 f.; c.f. HÄGELE, 2021, pp. 132). These romanticized and timeless images appeared, for instance, in songs and poetry which praised the beautiful *Heimat*. However, these praises often depicted generic and homogenous images, thus the extolled *Heimat* was replaceable (BAUSINGER, 1986, p. 96). BAUSINGER describes the bourgeois concept of *Heimat* as a space of appeasement [*Besänftigungslandschaft*], which compensated the tensions of the reality. Thus, *Heimat* functioned as a compensational space [*Kompensationsraum*], that offered a pleasurable and attractive 'world of strolling' [*Spazierwelt*] (1986, p. 96).

---

[3] The concept of *Heimweh* [homesickness] appears for the first time in a dissertation titled *De Nostalgia, Oder Heimwehe* by the medical student Johannes Hofer presented in Switzerland in 1688. As the term *Heimweh* seemed to be inappropriate for a dissertation, he coined the word *nostalgia* (TUAN, 2012, p. 232). Here, *nostalgia* or *Heimweh* is understood as a strong and painful longing to return to one's *Heimat*, which can lead to such psychopathological symptoms as depression or a loss of appetite, and therefore is considered as a disease for a long time (BLICKLE, 2002, pp. 67 f.). "Until the second half of the eighteenth century *Heimweh* was so much associated with Swiss mercenaries, who were sent all over Europe to fight, that it was often referred to as *die Schweizer Krankheit* [the Swiss Disease]" (BLICKLE, 2002, pp. 67 f.).

Amplified by the internal political crisis situation after 1890, the concept of *Heimat* experienced an ideologization (c.f. BASTIAN, 1995, p. 122). At the turn of the century, the concept of *Heimat* was specifically allocated to rural life, which was glorified by the so-called *Heimat(schutz)bewegung* [*Heimat* (protection) movement]. The *Heimat(schutz)bewegung* comprised, for the most part, the bourgeoisie as well as public servants (ROLLINS, 1996, p. 93) and a few farmers who opposed modernism and the political structures of Wilhelmine Germany, in particular the new economic policy, industrialization, urban lifestyle, internationalism, individualism and consequences such as alienation and environmental degradation. The movement aimed to conserve nature as well as tradition and communion as a patriotic virtue (BASTIAN, 1995, pp. 122 f.; c.f. KÖSTLIN, 2010, p. 27; HÄGELE, 2021, p. 134).[4] *Heimat*—in the understanding of the movement—was constructed around "[...] bourgeois ideals of family, class, gender roles, history, and politics" (BLICKLE, 2002, p. 20). The German culture and media scholar Ulrich HÄGELE remarks that *Heimat* presented a "set of ethical and moral standards of value, all forming part of a strategy to seek to neutralise elements of anything considered 'foreign' and 'modern'" (2021, p. 132). To achieve their past oriented political objectives, adherents of the movement founded, on the one hand, platforms such as the Life Reform Movement [*Lebensreformbewegung*] and the Conservation Movement [*Naturschutzbewegung*] (HÄGELE, 2021, p. 132) as well as *Heimat* associations [*Heimatvereinigungen/Heimatbünde*] in all regions in order to educate public opinion (ROLLINS, 1996, p. 95) through, for instance, the establishment of *Heimat* museums and *Heimatkunde* as a new subject with local focus in schools. On the other hand, they also initiated an artistic *Heimat* movement [*Heimatkunstbewegung*] to emphasize their values, primarily in the form of literature which depicted ideal, harmonic, and heroic village communities in the past (BAUSINGER, 1986, p. 100). BAUSINGER notes that *Heimat*, already at that time, became more and more a commercialized backdrop of the culture industry, as it was reduced to single stereotypical elements of German-language cultures in order to create an atmosphere of comfort and longing (1986, pp. 102 f.). Stereotypical representations, simplifications, and the lack of detail indicate that *Heimat* was not about real places (ROLLINS, 1996, p. 97). Images of a rural *Heimat* rather reflected a longing for security and familiarity in the face of growing uncertainty and anonymity (c.f. JOISTEN, 2003, p. 20). HÄGELE notes that visual representations

---

[4] The concept of *Heimat* of the *Heimat(schutz)bewegung* differs fundamentally from the understanding of the labor movement [*Arbeiterbewegung*], where the movement itself presents *Heimat* for the laborers. Their understanding of *Heimat* is characterized by (international) solidarity, and thus it is not limited to a specific place, but rather it is localized within a community. Thus, the proletarian *Heimat* is not a conservative and static idea, but a forward-oriented aspiration (BAUSINGER, 1986, p. 99; c.f. BASTIAN, 1995, pp. 125 f.).

of *Heimat* in the form of photographs [*Heimatfotografie*] of local costumes, architecture, and handicraft, for instance, served to preserve traditional culture of the rural world. He emphasizes, however, that photography did not only function as a tool for preservation, but rather as a medium to create a unified identity by familiar images and thus as an iconography of the nation (2021, pp. 132 f.; c.f. HERRMANN, 1996, p. 9). Hence, *Heimat* did not only represent a longing for familiarity and identity, but it was also a vehicle to construct identity and community. Thereby, the regional and individual understanding of one's *Heimat* shifted through unified narratives and images to a collective idea of *Heimat* and a collective identity, which is seemingly built on a historic continuity of common traditions of one nation. The concept of *Heimat* was thus central for building a national identity and identifying with a national territory, the fatherland [*Vaterland*], which needed to be preserved and protected (BASTIAN, 1995, pp. 123 f.; c.f. ROLLINS, 1996, p. 98).

Total-connectedness between Germanness and soil
After the First World War, the understanding of *Heimat* was again characterized by compensation with enhanced ideological, mystic, and utopian qualities (JOISTEN, 2003, p. 20).

> "During the 1920s, a discourse emerged that was focussed on the evolution of German society and this gathered momentum, not least because of the lost war. The question of the origins and capacities of the people themselves became a central theme of this inquiry in order to determine how an individual's provenance affected their contribution to the modern era" (HÄGELE, 2021, pp. 134 ff.).

The National Socialists reinforced pre-existing ideological charges of the concept of *Heimat* and further enhanced narratives of the German *Volk* as a unified natural community [*Volksgemeinschaft*], which is rooted in 'mother soil' (HÄGELE, 2021, p. 139). The German Nation, territory and soil became exaggerated and were considered as the *Ur-Heimat* [original *Heimat*] of the German *Volk*. This exaggeration expressed, on the one hand, in the alignment of the emotional value of *Heimat* with terms such as 'community', 'nation', '*Volk*' and 'German' (BASTIAN, 1995, p. 135), which was important with regard to the identification of the population with the rather abstract construct of a fatherland as well as the mobilization of the population in order to defend the victimized *Heimat* and to support the Second World War (JOISTEN, 2003, p. 21; c.f. BLICKLE, 2002, p. 19). And on the other hand, this exaggeration expressed in a total connectedness [*Totalverbundenheit*] with soil, the so-called blood-and-soil-ideology (BASTIAN, 1995, pp. 132). In this ideology "[...] *Heimat* does not just "belong" to Germans; Germans *belong* to *Heimat*"

(MITCHELL, 2000, p. 265). Who was considered as German and thus as belonging to this ethnic idea of *Heimat* was defined by the ideology of "contemporary anthropological-medical racial hygienists" (HÄGELE, 2021, p. 144). Disabled and homosexual people, as well as Jews, Sinti, Roma, and other racialized people were considered as foreign and 'the other', and thus were excluded from Germanness as well as *Heimat*. This was exemplified, for instance, in medial representations, such as magazines, which were concerned with moralizing notions of purity, 'cultural harmony' and worthiness (HÄGELE, 2021, pp. 145; c.f. MITCHELL, 2000, p. 265). If 'the other' was represented at all, images "[...] were visually staged and provided with captions that rejected and often demeaned the subjects" (HÄGELE, 2021, p. 144). Hence, 'the other' only appeared in stigmatizing and racial-ideological contexts (HÄGELE, 2021, p. 160). In contrast, National Socialists idealized homogenous large peasant families with blond hair in traditional costumes (HÄGELE, 2021, p. 149). According to HÄGELE, the understanding of *Heimat* propagated through media during National Socialism can be best explained by what is not shown: "[Socially] critical themes and day-to-day life, industrialization, and armaments production are as absent as intercultural coexistence in the city and in the countryside" (HÄGELE, 2021, p. 160) and "[nor] did they address those groups of the population that increasingly suffered from racial ideology" (HÄGELE, 2021, p. 169). A further characteristic of the national-socialist idea of *Heimat* was the expansionist ambitions based on the concept of *Lebensraum* [living space] by German ethnographer and geographer Friedrich Ratzel (c.f. SANDLER, 2012). This reflects in the terms 'Third Reich' and 'Greater German Reich' [*Großdeutsches Reich*] (BASTIAN, 1995, p. 133). However, the national imagination and expansionist ambitions of the National Socialist were not free of contradictions, because "[...] national unity had different regional interpretations and instrumentalizations, whereby the tensions between *Volksgemeinschaft* and *Heimat* became apparent" (CHU, 2012, p. 73; c.f. BAUSINGER, 1986, p. 105). Moreover, the social, economic, and transport policy of National Socialism with a focus on mass consumption, modernization, and mobility countered the seemingly idealized rural life and propagated image of *Heimat* (SCHARNOWSKI, 2019, p. 98).

A Janus-faced concept: Rejection and rediscovery of *Heimat*
After the Second World War, the concept of *Heimat* in the Federal Republic of Germany needed to be redefined and thus reverted from a Great German Reich to a local understanding of *Heimat*.[5] To a certain extent, *Heimatschlager* [*Heimat*

---

[5] In the socialist system of the German Democratic Republic the concept of *Heimat* received rather little attention. It was merely used as a political tool to create a collective awareness in

songs], *Heimatfilme* [*Heimat* films] and *Heimatromane* [*Heimat* literature] accelerated a depoliticization of *Heimat*, as they produced a pleasant atmosphere by depicting idyllic (mountain) landscapes and dramatic stories of rural communities (c.f. COSTADURA et al., 2019b, p 14). Nevertheless, the notion of *Heimat* initially received great skepticism and rejection, mostly among younger generations, as it was perceived as a retrograde concept, which is connected to ideological values of the previous generation and to bourgeois parochialism (BASTIAN, 1995, pp. 139 f.; c.f. BAUSINGER, 1986, p. 107; WEBER, 2019, p. 7). *Heimat* in a political sense especially "[...] continued to play a role as part of a nostalgic and backwards-looking milieu, even if only from the point of view of those who were part of the millions of displaced Germans [*Heimatvertriebene*] who had fled or been forcibly expelled from former German territories" (HÄGELE, 2021, p. 170; c.f. JOISTEN, 2003, p. 21; WEBER et al., 2019, p. 7). However, in the 1970s, the ideological preload of *Heimat* seemed to be overcome by a redefined political understanding of the concept (COSTADURA et al., 2019b, p. 14). The turn to regionalism and a growing ecological awareness due to the increase in landscape destruction and environmental degradation strengthened the significance of local identity, a regional understanding of *Heimat* and an awareness for the qualities of *Heimat*. Due to this awareness, a wide-ranging politically and socially diverse protest movement developed on the shared motivation to oppose envisaged projects which destroy landscapes or endanger the environment, such as nuclear power plants, highways, landfill sites and so on (BASTIAN, 1995, pp. 140 f.). In other words, the critical stance towards economic growth and the drawbacks of globalization and modernity led to the rediscovery of *Heimat* (JOISTEN, 2003, p. 21; c.f. BAUSINGER, 1986, pp. 108 f.). Here, the political concept of *Heimat* reveals its ambiguous qualities as, on the one hand, it can be considered as a retrospective and nostalgic idea, but on the other hand, it can be viewed as a forward-oriented motive to actively engage in local political, social, and ecological matters. Thus, *Heimat* needs to be actively created (c.f. JOISTEN, 2003, p. 22; BAUSINGER, 1986, p. 109). However, in the years after the reunification of Germany, *Heimat* once again became an emotive term with regard to the convergence of Eastern and Western Germany and efforts of integration. Due to its identity building qualities, the concept of *Heimat* was especially popular among conservative political actors (BASTIAN, 1995, p. 143). Rising insecurities over German identity and the economic development, hence, also lead, at least to some degree, to increasing negative attitudes and violence against immigrants, especially against so-called

---

the sense of socialist ideology. The socialist *Heimat* was linked to the fight against capitalism, and thus was perceived as a part of the global socialist system and the belief in social and political progress (BASTIAN, 1995, pp. 136 ff.).

'guest workers' [*Gastarbeiter*innen*] or asylum-seekers, which were perceived as foreigners [*Ausländer*innen*] (MITCHELL, 2000, pp. 264 f.). Once again, the concept of *Heimat* reveals its exclusionary potentials, due to national and ethnic definitions of *Heimat* (c.f. COSTADURA et al., 2019b, p. 17; BLICKLE, 2002, p. 159).

Conclusions from the history of the *Heimat* concept

Although this brief overview cannot trace every aspect of the historical shifts and uses of the concept of *Heimat* in German language cultures, some general conclusions can be drawn. Like a 'chameleon' (BAUSINGER, 2009), the notion of *Heimat* shifted its meaning significantly over the years (HÄGELE, 2021, p. 131). As *Heimat* occurs wave-like in public discourse (WEBER et al., 2019, p. 7), it can be regarded as a "seismograph for social upheavals" (BÖNISCH et al., trans., 2020b, p. 7) or as an indicator of the conditions of specific times in history. Notions of *Heimat* are emotional and subjective but also collective, social, and political. The historical concept poses a polyvalent amalgamation of questions around identity, belonging, history, politics, nature, and space (BLICKLE, 2002, p. 8). Hence, *Heimat* is a variable signifier as the concept "[…] fulfills different functions in different contexts" (BLICKLE, 2002, p. 77) and depends on the ideas and intentions of different actors (BASTIAN, 1995, p. 221). Historical impulses of the concept of *Heimat* still have an impact on present times (BAUSINGER, 1986, p. 107). As a result, the German idea of *Heimat* is multifaceted and characterized by ambiguities—it is backward and forward looking at once (c.f. JOISTEN, 2003, p. 22). Past oriented notions of *Heimat* foster nostalgic and romanticized imaginations, which often serve to express discomfort with modernity and to create a sense of coherence (c.f. KÖSTLIN, 2010, p. 36). As a consequence, the notion is also used to insist on rootedness and to express a fear of loss as well as to neglect and actively exclude certain subjects, which are perceived as being 'out of place' (c.f. HÄGELE, 2021, p. 145; MITCHELL, 2000, p. 269). Future-oriented notions of *Heimat*, however, may also encourage a critical and meaningful engagement with local environmental, social, or economic issues, which can contribute to active democratic participation (c.f. BAUSINGER, 1986, p. 111). Today the notion of *Heimat* is anchored in everyday life in German-language cultures and is used differently depending on references, such as space, time, identity, belonging and so on (WEBER et al., 2019, pp. 14 f.).

1.2   A Multifaceted Concept

As "[...] the notion of *Heimat* carries with it a long history of usages and appropriations, [...] [it has] accumulated multiple connotations that have turned the
concept into a rich reservoir for scholars in the social sciences and the human-
ities" (EIGLER and KUGELE, 2012b, p. 1). Nevertheless, due to this ambiguities
and complexities *Heimat* presents a challenge for scholarly attempts at defining the term (BLICKLE, 2002, p. 5; c.f. JOISTEN, 2003, p. 18). Consequently, the
notion of *Heimat* does not appear in sociological or ethnographic studies for
decades (BAUSINGER, 1986, p. 90). One of the first and groundbreaking studies,
and thus often cited, is *Der territoriale Mensch: Ein literaturanthropologischer
Versuch zum Heimatphänomen* [Territorial Human: An Attempt at Understanding
the Heimat Phenomenon Through an Anthropology of Literature] by the German
cultural anthropologist Ina-Maria GREVERUS (1972). She enriches the discussions
of *Heimat* by equating the term with territory or territoriality, whereby she refers
to a space of possession and defensiveness as *Heimat* represents the fulfillment
of primary human needs for identity, activity, safety, and satisfaction (BASTIAN,
1995, p. 37). In doing so, she shows how the idea of *Heimat* is related to German nationalism as well as antagonisms of the nineteenth and twentieth centuries
(BLICKLE, 2002, p. 10). Another important contribution to the discussions of
Heimat is made by the German cultural anthropologist Hermann BAUSINGER,
who traces the contradictions, transformations, and appropriations alongside the
history of the term's usage in his essay *Heimat in einer offenen Gesellschaft:
Begriffsgeschichte als Problemgeschichte* [Heimat in an Open Society: History
of a Concept as Problematic History] (1986). His explanation demonstrates the
concept's indicative and compensatory character as notions of *Heimat* increasingly appear in times of uncertainty or endangerment of rural life and landscape
due to modernization processes (c.f. BAUSINGER, 1986, pp. 100 ff.). BAUSINGER
emphasizes, however, also that *Heimat* requires active appropriation as it is a
"medium and objective of practical engagement" (trans., 1986, p. 111). In *Der
Heimat-Begriff: Eine begriffsgeschichtliche Untersuchung in verschiedenen Funktionsbereichen der deutschen Sprache* [The Concept of Heimat: A Study of the
History of a Concept in Different Contexts of the German Language] the German
linguist Andrea BASTIAN (1995) offers an even more extensive exploration of the
historical uses and shifts of the term in different contexts, such as everyday life,
law, politics, natural sciences, religion, and literature.

The above-mentioned influential studies demonstrate that the issue of *Heimat*
was initially primarily approached by scholars from the fields of cultural anthropology and literature studies (c.f. BÖNISCH et al., 2020b, p. 2). Over time,

however, diverse perspectives across the academic disciplines have resulted in a vast body of literature on different but also interrelated dimensions of the multifaceted concept of *Heimat*. Psychologist Beate MITZSCHERLICH (1997), for example, determines in her dissertation ten different dimensions of *Heimat* by conducting qualitative interviews. According to her, the contents of subjective *Heimat* conceptions can be summarized as follows: 1. *Heimat* as an environment of childhood, 2. *Heimat* as cultural landscape, 3. *Heimat* in current relationships, 4. *Heimat* as experience and emotional state, 5. *Heimat* as an inner concept, 6. *Heimat* as a political and ideological construction, 7. *Heimat* as folkloristic world, 8. *Heimat* as an experience of loss, 9. *Heimat* and *Fremde*, 10. *Heimat* and diversity (MITZSCHERLICH, 1997, p. 56). Further research from other perspectives includes monographies as well as anthologies[6] on the history of the concept in German-speaking contexts (e.g. HERMAND and STEAKLEY, 1996; WICKHAM, 1999; BOA and PALFREYMAN, 2000; BLICKLE, 2002; SZEJNMANN and UMBACH, 2012), the connection to landscape and nature conservation (e.g. FRANKE, 2017), the concept's imaginative and emotional qualities as a place of longing (e.g. WICKHAM, 1999; EIGLER and KUGELE, 2012a; EGGER, 2014), its importance regarding identity and a sense of belonging (e.g. BAUSINGER and KÖSTLIN, 1980; MOOSMANN, 1980; HERMAND and STEAKLEY, 1996), representations and discourses in media (e.g. VON MOLTKE, 2005; EIGLER and KUGELE, 2012a; EICHMANNS and FRANKE, 2013), the concept in the context of globalization, migration and mobility (e.g. EICHMANNS and FRANKE, 2013; HASSE, 2018; COSTADURA et al., 2019a; KÜCK, 2021; AL- ALI and KOSER, 2002a), and its contested position (e.g. BELSCHNER et al., 1995; BÖNISCH et al., 2020a). Despite the importance of spatial dimensions in the concept of *Heimat*, there are only a few explicit geographical compilations (e.g. HÜLZ et al., 2019). Nonetheless, a clear demarcation between disciplinary approaches and dimensions of the concept of *Heimat* is rarely possible. Many studies and anthologies attempt to encompass the multiplicity of the concept whether by approaching many different dimensions or by multi- and interdisciplinary approaches from cultural anthropology, literature studies, sociology, history, psychology, geography, or theology (e.g. GEBHARD et al., 2007; DONIG et al., 2009; SEIFERT, 2010a; KLOSE, 2013; BRINKMANN and HAMMANN, 2019).

---

[6] At this point only a selection of the literature on *Heimat* in the form of relevant monographies and anthologies can be presented since a comprehensive listing including single articles would go beyond the scope of this study.

1.3   Disciplinary Orientation and Objective of this Study

As the literature on *Heimat* presents the concept's inherent intersections of space, place, identity, belonging, and emotions depending on time and scale, the concept reveals its values especially for geographical research. Although *Heimat* is not inevitably linked to a certain place, but rather to social relations, emotions, experiences, and memories (WEBER et al., 2019, p. 12), the concept is inherently spatial. Following the categorization by BÖNISCH et al. (2020b), political, social, and cultural negotiations of *Heimat* are expressed in three different but also overlapping spatial semantics: "A reactionary and closed understanding equates 'Heimat' with 'nation'; another concept of Heimat is connoted by local and regional belonging; and fluid, processual conceptions of Heimat are either deterritorialized, decidedly transnational situated or emphasize Heimat(en) as practice" (BÖNISCH et al., trans., 2020b, p. 6). The intention of the present study is to argue along the line of the third spatial semantic, towards dynamic, plural, and performative conceptions of *Heimat*. The present study follows the argumentation that the multiple dimensions of *Heimat* are not exclusively embedded in the local and national, but also in global processes and reciprocal effects. Political, social, and cultural semantics and qualities of the concept can be translated into other contexts beyond German-language cultures. Thus, *Heimat* can be understood as a global phenomenon (BÖNISCH et al., 2020b, p. 7). Such progressive conceptions of *Heimat*, which pose a challenge to static conceptions, are the result of processes of fragmentation and pluralization due to the increase of national as well as international mobility and multilocality (WEBER et al., 2019, p. 12; c.f. COSTADURA et al., 2019b, p. 35; BLICKLE, 2002, p. 153). The cultural scholars Kathrin LEHNERT and Barbara LEMBERGER (2013) argue that mobility should not be considered as an opposite pole to settledness. Rather, the focus should be on processes of migration and practices of plurilocality (EGGER, 2020, p. 30). In order to gain insights on dynamic, plural, and performative conceptions, and hence to "demystify ideological Heimat conceptions" (MITZSCHERLICH, 2019, p. 184), which promote an exclusive and static understanding of the notion, it is worth examining *Heimat* through the lens of migration on an empirical level. In this sense, the present study refers to postmigrant approaches (c.f. YILDIZ and HILL, 2014a) and the perspective of transnational migration (c.f. FAIST, 2012; GLICK SCHILLER, 2007) in order to utilize migration as a perspective on the concept of *Heimat*. Accordingly, the theoretical part of this study is embedded in the discipline of globalization geography with a focus on identity, belonging and migration in the context of *Heimat* conceptions. Moreover, the theoretical framework also feeds from the approaches of the field of emotional geography, as experiences, emotions and memories constitute an integral part of the concept of *Heimat*. The focus of the empirical section of this study is on the plurilocal practices, experiences, and perceptions of international migrants, who have shifted their place of residence and thus their center of life to another country on a long-term basis. The aim is to gain insights into individual senses of *Heimat* with respect to trans- and plurilocal practices and the intersections of space, time, identity, belonging, and emotions. These insights are to be obtained from conducting biographical interviews with people who have had experiences of international migration and settling into an environment that differs from the places in the country that was left in terms of language, daily social practices, and other elements that constitute a familiar space. The objective of the present study is also to demonstrate that the dimensions and connotations of *Heimat* are universal, regardless of its roots in the German language (c.f. BÖNISCH et al., 2020b, p. 7). In order to gain an insight on the relevance, multifaceted dimensions and development of the sense of *Heimat* in a globalized world, the following central research questions are phrased: *In how far does international migration challenge static conceptions of Heimat? How are the biographies of international migrants reflected in their sense of Heimat?* 'Sense' in 'sense of *Heimat*' is deliberately conceived in a double meaning in the present thesis. On the one hand, as the individual conception of the term—i.e. as *Heimat* is understood by individuals—and, on the other hand, as the emotional level of the phenomenon and space-producing feeling.

This study is structured as follows: Chapter 2 will present the theoretical framework in order to understand how *Heimat* is conceptualized in this study. In Section 2.1 *Heimat* will be considered as an inherently spatial conception since it is a product of multidimensional processes of construction. This constructivist approach will be discussed in more detail in Section 2.1.1, which will explore the role of experiences, memories, emotions and feelings in the development of an individual sense of *Heimat*. Section 2.1.2 will pay particular attention to spatial negotiations of identity and belonging in the context of *Heimat* and practices of self-location. In Section 2.2 *Heimat* will be contextualized within a globalized world. For this purpose, Section 2.2.1 will discuss the field of tensions between stability and change, which refigurates locally conceptualized spaces such as *Heimat* and can be considered as the result of increased mobility and flows of migration. Subsequently, Section 2.2.2 will focus on a plurilocal sense of *Heimat* in the context of international migration, translocal living realities and practices in order to reconsider a static and monolithic understanding of *Heimat*. To complete the theoretical part of this study, Section 2.3 will provide a summary of the theoretical framework in the form of an operational definition of *Heimat*, which provides the foundation for the methodological approach and poses the framework for the analytical part of this study. Chapter 3 presents the methodological approach, which includes the selection of participants, the implementation of the qualitative data collection methods of biographical online interviews and solicited online diaries, as well as the data analysis method of qualitative content analysis. The results of this study are then presented in Chapter 4 and discussed in Chapter 5 in relation to the preliminary theoretical considerations and the research questions. Finally, a conclusion is drawn in chapter 6.

Heimat: Theoretical Framework(s)   2

Since *Heimat* constitutes a polyvalent concept, which can be approached from multiple perspectives, it is necessary to elaborate a theoretical framework in order to distill an operational definition. For this purpose, *Heimat* will first be subjected to its spatial dimensions on a conceptual level (Section 2.1) and, subsequently, will be contextualized in a globalized world (Section 2.2). Finally, the preceding theoretical considerations will be merged into an operational definition of *Heimat* (Section 2.3).

2.1   *Heimat*: a Spatial Concept

In order to examine the individual understanding and everyday significance as well as the more general qualities of the concept, it is worth examining *Heimat* from a geographical perspective. Although *Heimat* is often linked to spatial notions, spatial theory was neglected in scientific research on *Heimat* for a long time (RUNIA, 2020, p. 167). The article *Critical Approaches to Heimat and the 'Spatial Turn'* by literary scholar Friederike EIGLER (2012) strongly contributed to an increased attention to spatial theory in relation to conceptions of *Heimat* (RUNIA, 2020, p. 167). EIGLER notes that critical approaches within German studies examined, in response to ideological or reactionary manifestations of *Heimat*, specific representations of space and place under special consideration of history and politics (2012, pp. 37 f.). She critically remarks that "[t]hey may thus have inadvertently contributed to what looks, from the perspective of cultural geography, like a rather static, fixed notion of place" (EIGLER, 2012, p. 38). The essentialization of fixed notions of place often also reflects in scientific as well as in non-scientific literature, which either claim that *Heimat* is not a place (c.f. LESZCZYNSKA-KOENEN, 2019) or point out that *Heimat* cannot be reduced to

J. Andel, *Sense(s) of Heimat*, BestMasters,
https://doi.org/10.1007/978-3-658-38985-7_2

spatial references like birthplace, place of residence, region, or landscape. Other factors such as social relations, societal structures as well as memories, experiences and emotions play an equally important, if not a more decisive role in the individual sense of *Heimat* (c.f. BASTIAN, 1995, pp. 40 ff.; SCHRAMM and LIEBERS, 2019). It is, of course, indisputable that these aspects are significant for a sense of *Heimat*, but this argumentation is questionable. On the one hand, social relations, experiences, and memories that are constitutive for a sense of *Heimat* always 'take place'—in the literal sense of the phrase. *Heimat* is not always tied to a specific place, but a sense of *Heimat* is oriented towards social relations, emotions, experiences, and memories which are often connected to certain places. And on the other hand, this argumentation is founded, from a cultural geographic perspective, on the traditional and unreflected understanding of space as a container in which the physical-material world and anthropogenic factors stand in a mutual relationship (c.f. WARDENGA, 2002, p. 47). Space appears to be understood as fixed to its physical-material manifestations. However, "[i]n the broader sense, Heimat can be understood as an affective relation between human and space—whereat [this space] can be of geographical, cultural, or also social nature" (BÖNISCH et al., trans., 2020b, p. 2). Space should be considered as a social category, which is determined by social interaction, relations, interdependency and processuality. Therefore, space can be regarded as the medium of human action (LÖW and KNOBLAUCH, 2021, p. 27; BRUNS and MÜNDERLEIN, 2019, pp. 104 f.). If *Heimat* is thus also considered as an open subjective construct which is a product of multidimensional processes and social practices (BÖNISCH et al., 2020b, p. 8; c.f. EIGLER, 2012; MASSEY, 1995), then *Heimat* is an inherently spatial conception. "Thus, Heimat is an interactive space, which is continuously changing and emerges from interaction—interaction by humans, by discourses, by media, by knowledge etc." (COSTADURA et al., trans., 2019b, p. 22). In other words, *Heimat* constitutes a synthesis of human spaces of action and imagination (SEIFERT, 2010b, p. 20).

In the following sections, *Heimat* will be conceptualized as a spatial construct, which emerges from the subjective experience of space. As emotions, feelings, memories and other cognitive processes play a crucial role in the development of a sense of *Heimat*, it is beneficial for this study to examine approaches from the field of emotional geography (Section 2.1.1). Furthermore, the constitutive aspect of (inter)actions indicates the importance of (social) practices. In particular with regard to identity and a sense of belonging it is envisaged to draft *Heimat* as a practice, which can be understood as a practice of self-location (Section 2.1.2).

2.1.1 *Heimat*: an Emotional Geography

Since the geographers Kay ANDERSON and Susan J. SMITH "[...] have been reflecting on the extent to which the human world is constructed and lived through emotions" (2001, p. 7) and called in their editorial paper for "[...] an awareness of how emotional relations shape society and space [...]" (2001, p. 9), the importance of affects, feelings and emotions received great appreciation in geographical research (PILE, 2010, p. 6). The study of emotions within geography is particularly relevant, because emotions can be understood as dynamic processes that enable the individual to experience and to interpret a changing world, as well as to position oneself in relation to other individuals, and to shape one's own subjectivity (SVAŠEK, 2010, p. 868). According to the geographer Liz BONDI, the concern with emotions, feelings and affects can be traced back to "[...] at least three pre-existing and sometimes overlapping geographical traditions, namely humanistic geography, feminist geography and non-representational geography [...]" (2005, p. 435). In humanistic geography the research interest is directed towards the subjective experience of space and place (SCHURR, 2014, p. 149). Only through experience, which is the totality of sensation, perception, and conception, people are able to understand their environment (TUAN, 2016 [1974], p. 133). However, it is important to note that "[t]he structure and feeling-tone of space is tied to the perceptual equipment, experience, mood, and purpose of the human individual. We get to know the world through the possibilities and limitations of our senses" (TUAN, 2016 [1974], p. 142). As *Heimat* is often described as an individual feeling, it is thus worth examining *Heimat* from the perspective of humanistic geography. In this context, reference is often made to the work of the geographer Yi-Fu TUAN (2016 [1974]; 1977). He argues that social relations and affective attachment transform abstract space into place (EIGLER, 2012, p. 36; BRUNS and MÜNDERLEIN, 2019, p. 105). Individual experiences as well as the perception of physical characteristics of space lead to people "[...] apply[ing] their moral and aesthetic discernment to sites and locations" (TUAN, 2016[1974], p. 152). Through this 'sense of place' (TUAN, 2016 [1974], p. 152) a mere physical-geographical location converts "[...] into a place with special behavioral and emotional characteristics for individuals" (HASHEMNEZHAD et al., 2013, p. 5). TUAN considers place as an expression of the aspirations and experiences of people (2016 [1974], p. 133). According to him, "[a] key to the meaning of place lies in the expression that people use when they want to give it a sense carrying greater emotional charge than location or functional node" (2016 [1974], p. 151). As a result, people develop a certain sense of belonging to places in emotional terms (SCANNELL and GIFFORD, 2010, p. 3). The processes that contribute to the formation of a sense of *Heimat* can be summarized under the term 'place-making' (or in this case '*Heimat*-making'), which describes the socio-emotional appropriation of space (BRUNS and MÜNDERLEIN, 2019, p. 105). Following the argumentation of TUAN (2016 [1974]), defining *Heimat* as a feeling, thus, implies that the immaterial dimensions of *Heimat* derive from experiences which are connected to places. Emotions and memories, hence, cannot be examined detached from spatial references, as feelings are connected to and derive from them. It can be further argued, that from a constructivist perspective, those immaterialities manifest themselves in a space called *Heimat*.

A safe place

A connection between the emotional values of *Heimat* and space can already be found in the work of GREVERUS (1972; 1979). She uses the principle of territoriality to describe a space of identification, which also constitutes a space of security and action. This territory rests on familiarity (GREVERUS, 1972, p. 382). In this respect, familiarity does not only refer to being familiar with routes and locations, but also being familiar with values, norms and other expressions of social organization (MITZSCHERLICH, 2019, p. 187; c.f. BASTIAN, 1995, pp. 40 f.). One could also say that this space is characterized by certain regularities (c.f. DOUGLAS, 1991, p. 289). In reference to the book *Modernity and Self-Identity* by sociologist Anthony GIDDENS (1991a), BLICKLE writes that "[…] in Heimat conceptualizations an impersonal trust in systems and symbols is irrelevant because everything in this locality is known (*bekannt* and *gekannt*)" (2002, p. 32). *Heimat* is thus also a space of "knowing, being known and being acknowledged" (GREVERUS, trans., 1979, p. 13.). Similar to GREVERUS' explorations, MITZSCHERLICH (2019) identifies three dimensions of the *Heimat* feeling [*Heimatgefühl*]: 'Sense of Community', 'Sense of Control' and 'Sense of Coherence'. With 'Sense of Community' she refers to the experience of (social) belonging and involvement. Thereby she also refers to the experience of 'knowing, being known and being acknowledged'. 'Sense of Control' describes the experience of behavioral safety and capacity for action. This means, for example, being aware of how and what to do in order to accomplish a certain goal. With 'Sense of Coherence' she relates to the (subjective) interpretation of 'being in the right place' (MITZSCHERLICH, 2019, pp. 187 f.). Thus, *Heimat* is a place one can always come back to as it is the center of an individual's concern and control (RAPPORT and DAWSON, 1998b, p. 6).

Feelings of *Heimat* are, first of all, associated with the basic human desire for security and safety (COSTADURA et al., 2019b, p. 19; c.f. JOISTEN, 2003, p. 25; BASTIAN, 1995, pp. 33 f.). Consequently, *Heimat* is connected to dwelling (COSTADURA et al., 2019b, p. 18) and thus to the concept of home. However, the concept of home, like the concept of *Heimat*, is contested in academic literature. Home is regarded "[…] as a socio-spatial entity, a psycho-spatial entity and an emotional 'warehouse'" (EASTHOPE, 2004, p. 134). In the words of TUAN, "[h]ome is a place that offers security, familiarity and nurture" (2004, p. 164) and can be defined as a "[…] locale of human warmth and material sustenance, moral probity and spiritual comfort" (TUAN, 2012, p. 227). Hence, home can be considered as a safe place (TUAN, 2012, p. 226). This place does not only offer physical protection but also represents the foundation of personal and social meaning (PAPASTERGIADIS, 1998, p. 2). The geographer Alison BLUNT writes that

> "[t]he home is a material and affective space, shaped by everyday practices, lived experiences, social relations, memories and emotions, [however] […] the meanings and lived experiences of home are diverse, [as it also can be] […] a space of belonging and alienation, intimacy and violence, desire and fear […]" (2005, p. 506).

James DUNCAN and David LAMBERT conclude in their geographical study on *landscapes of home* that home is "[…] perhaps the most emotive of geographical concepts […]" (2003, p. 395). On account of the manifold parallels between the concept of home and *Heimat*, the notion of home is often used as an equivalent of the German concept of *Heimat*. Both concepts cannot be clearly distinguished from one another (JOISTEN, 2003, p. 19). The philosopher Karen JOISTEN emphasizes that *Heimat* is unthinkable without home as the place of dwelling (2003, p. 19). However, unlike *Heimat*, home is always localizable, although it is not necessarily fixed (DOUGLAS, 1991, p. 289). JOISTEN views *Heimat* as the spatial extension of home with all of its (positive) emotional connotations (2003, p. 19).

Formation of place attachment
The predominant positive emotional connotations may also result from the fact that *Heimat* is often associated with the places where individuals were born and spent their childhood and youth (c.f. JOISTEN, 2003, p. 25). Memories of childhood are, according to TUAN, "[t]he most vivid hauntings […] [because] children have no pressing responsibilities that divert them from full sensory experience" (2012, p. 228). He further writes that "[c]hildren apprehend space before time […] [b]ut the appreciation of historical time comes much later" (2016 [1974], p. 137). Places and, above all, the related memories of childhood are constitutive for an individual's sense of *Heimat*. The places of growing up and early socialization are charged with memories that orient the individual. "[…] [I]t is in these 'original' spaces and places that so many of the social roles and categories that define the self-identity of the person, are learned or internalized by the child […]" (PROSHANSKY et al., 1983, p. 64).

Furthermore, childhood experiences orient the individual, for instance, towards landscapes, which seem familiar (c.f. WEBER et al., 2019, p. 12; RISHBETH and POWELL, 2013) or 'homely', and towards familiar sensations, such as fragrances, sounds or tastes (c.f. TUAN, 2016 [1974], pp. 152 f.). Such childhood memories connected to places constitute meaningful environments and thus engender person-place bonds. Emotional bondings between people and the places that are important to them can be generally defined as place attachment—a concept which is occasionally used as a synonym for sense of place because of their similarities (BRUNS and MÜNDERLEIN, 2019, p. 106; c.f. HASHEMNEZHAD et al., 2013; SCANNELL and GIFFORD, 2010, p. 3). In general, place attachment can be considered as "[…] the pattern of reactions that a setting simulates for a person. These reactions are a product of both features of setting (what settings are) and personal processes (what the people bring to it) […]" (INALHAN and FINCH, 2004, p. 123).

The psychologists Leila SCANNELL and Robert GIFFORD identify three dimensions of place attachment from a variety of definitions in literature: the actor(s), the psychological process and the place as the object of attachment (2010, p. 1). The first dimension refers to both the individual and the group level. At the individual level, place attachment refers to personal meaningful connections to places based on personally important experiences. This not only refers to childhood experiences, but in general to 'in-place-experiences', such as personal growth, realizations or milestones, which create meaning and memories, and thus "[…] contribute to a stable sense of self" (SCANNELL and GIFFORD, 2010, p. 2; c.f. TWIGGER-ROSS and UZZELL, 1996). Collective place attachment "[…] is comprised of the symbolic meanings of a place that are shared among members [of a group]" (SCANNELL and GIFFORD, 2010, p. 2; c.f. LOW, 1992), which is based on certain experiences and other commonalties along supposed distinctions such as culture, nation, race, ethnicity, religion, gender, class, age and so on (PROSHANSKY et al., 1983, p. 64). On a smaller scale, collective attachment and meanings of places can also be shared among members of a family, a group of friends or other people from the social environment of an individual (c.f. ROBERTS, 2012). The second dimension, the psychological process, concerns affective, cognitive, and behavioral components. Affect, in this context, refers to an emotional connection to or investment in a particular place. Affective person-place relationships can be expressed in a variety of emotions from positive, such as love or satisfaction, to negative ones, such as fear, hatred or apathy (SCANNELL and GIFFORD, 2010, p. 3). Place attachment, however, is usually defined by positive emotions, as "[…] the desire to maintain closeness to a place is an attempt to experience the positive emotions that a place may evoke" (SCANNELL and GIFFORD, 2010, p. 3; c.f. GIULIANI, 2003). Cognitive components include memories, beliefs, meaning, and knowledge that facilitate person-place bonds. Place attachment is often based on memories, for instance "[...] representations of the past that the setting contains" (SCANNELL and GIFFORD, 2010, p. 3). TUAN writes that people can acquire a profound sense of place through time. With this he means that people can only be aware of their attachment to places, if they leave and view them from a distance (2016 [1974], p. 153). In other words, only when the sense of place is interrupted, one becomes aware of it (PROSHANSKY et al., 1983, p. 61; c.f. INALHAN and FINCH, 2004, p. 124; ROSE, 1995, p. 95). Reflection and appreciation of places is carried out through a process of distancing in the form of thinking and talking about these meaningful places (PROSHANSKY et al., 1983, p. 61). Through memory, people not only create meaning, but also incorporate information into their self-concept. This incorporation expresses itself, for example, in the attachment to certain types of places, as they seem familiar to the individual (SCANNELL and GIFFORD, 2010, p. 3.). However, through the process of remembering, actual experiences of an individual become "[...] modified by the cognitive process of memory and interpretation and such others as fantasy and imagination" (PROSHANSKY et al., 1983, p. 62). Behavior, in this context, refers to the way how the attachment to a place is expressed through actions. Place attachment is, for example, often characterized by 'proximity-maintaining behaviors', which are "[...] founded on the desire to remain close to a place, and can be expressed in part, by [...] place reconstruction, and relocation to similar places" (SCANNELL and GIFFORD, 2010, p. 4). The third dimension of place attachment, which SCANNELL and GIFFORD consider as the most important one, is the place itself with its various spatial scales (home, neighborhood, town, etc.). Place, in this context, can be divided into social and physical aspects, whereat both social and physical attachment influence the person-place bond (c.f. HIDALGO and HERNÁNDEZ, 2001). SCANNELL and GIFFORD note that "[...] much of the research on place attachment (and related concepts) has focused on its social aspects; people are attached to places that facilitate social relationships and group identity" (2010, p. 4). This is because individuals often perceive physical settings as 'backdrops' of social events and thus tend not to be aware of physical features and psychological processes which build and shape their connection to places (PROSHANSKY et al., 1983, p. 63). Thus, being attached to a place implies that people are attached to other people (family, friends, romantic relationship etc.), settings of social gatherings (educational institution, religious community, sports club etc.), and to the social interactions within that place, which foster a sense of belonging and familiarity (SCANNELL and GIFFORD, 2010, pp. 4 f.; MITZSCHERLICH, 2019, pp. 186 f.). Physical features of places, such as the aesthetics of the 'natural world' or climate, however, also influence the attachment of individuals, especially when they resemble physical features of one's childhood, as they represent one's past (SCANNELL and GIFFORD, 2010, p. 5).

According to SCANNELL and GIFFORD place attachment exists because it serves several functions. First of all, places offer survival advantages. Besides the provision of the necessities such as food, water and shelter, the attachment to places supports the need for satisfaction, security and safety (2010, pp. 5). This "[…] includes familiarity of place, a sense of community support, a sense of belonging and feeling of permanence" (INALHAN and FINCH, 2004, p. 125). Moreover, the attachment to a place forms when it supports the attainment of goals through the provision of required resources (SCANNELL and GIFFORD, 2010, p. 6). Thus, place attachment facilitates a sense of control, self-determination and the ability to initiate change and regulate interactions (INALHAN and FINCH, 2004, p. 125). Other functions of place attachment include the provision of identity building and enhancing factors (SCANNELL and GIFFORD, 2010, p. 6). The tripartite organizing framework and functions of place attachment, presented by SCANNELL and GIFFORD (2010), align with the explorations on *Heimat* by GREVERUS (1972) as well as the dimensions of the *Heimatgefühl* identified by MITZSCHERLICH (2019). These similarities make the concept of place attachment useful to understand the concept of *Heimat* and its emotional dimensions.

2.1.2 *Heimat*: a Practice of Self-Location

It is evident from the German history of the notion of *Heimat* as well as from the previous section about *Heimat* as an emotional geography that the concept is strongly linked to identity and a sense of belonging. *Heimat*, thus, is inevitably connected to the conception of the self. In general, the self can be considered "[…] as a term which describes the individual as a total system including both conscious and unconscious perceptions of his [or her] past, his [or her] daily experiences and behaviors, and his [or her] future aspirations" (PROSHANSKY et al., 1983, p. 58). In other words, it is "[…] a multiple, relational being-in-the-world that is captured by his or her surroundings, engaging with past, present and future situations" (SVAŠEK, 2010, p. 868). This being-in-the-world is the result of the complex interplay between biological factors and social-environmental influences. The self is expressed, for example, in a person's characteristics, such as speech melody, physical expression, taste preferences, and so on (CONRADSON and MCKAY, 2007, p. 167). The individual needs to have some sort of self-definition or sense of identity as it is fundamental for life in human society (BLICKLE, 2002, p. 64). As the concept of identity is often used in an inflationary manner and insinuates a continuity in life, the ethnologist Irene GÖTZ (2010) pleads for certain premises in order to use and understand self-conceptions. She suggests, firstly, that identity is used as a term for everyday life experiences of individuals and it refers in this context to the internal view and self-image of social actors. Secondly, identity can also be viewed from an external perspective, as it is a result of cultural practices (e.g. speaking) and reciprocal processes of identification with groups (GÖTZ, 2010, p. 206). In order to define the self and to identify with certain groups, the individual has to identify 'the other'. Thus, identities are positional (EASTHOPE, 2009, p. 68). This means, that "[...] people's identities are in part constituted by their definitions of what they are not and by the creation of (physical and mental) borders or boundaries around their identities" (EASTHOPE, 2009, p. 68). And thirdly, GÖTZ emphasizes that societal and personal crises or key experiences often lead to an increased awareness and thematization of identity (2010, p. 206). Further, she also adds the crucial premise that identities have to be considered as constructed, plural and context-dependent (GÖTZ, 2010, p. 210).

The link between self-conception and *Heimat* is also confirmed by BLICKLE, who writes that "[t]he interrelationship between a sense of self and the perception of one's world is a central aspect in the formation of a sense of Heimat" (2002, p. 61). Furthermore, he notes that *Heimat* is not simply a trope of identity, but rather a manifestation of identity, as "[i]t is a way of organizing space and time and a communally defined self in order to shape meaning" (2002, p. 66). Following this argumentation, *Heimat* can be understood as a sense of spatial self. Here, again, spatial references play a crucial role, as self-definitions derive from them (c.f. BASTIAN, 1995, p. 38). "This occurs when individuals draw similarities between self and place, and incorporate cognitions about the physical environment (memories, thoughts, values, preferences, categorizations) into their self-definitions" (SCANNELL and GIFFORD, 2010, p. 3). The environmental psychologist Harold M. PROSHANSKY and his colleagues (e.g. PROSHANSKY, 1978; PROSHANSKY et al., 1983; PROSHANSKY and FABIAN, 1987) introduce the term 'place identity'—which is regarded as an aspect of the concept of sense of place (LENGEN, 2019, p. 123)—in order to describe the role of 'physical world socialization' in the development of self-identity (PROSHANSKY et al., 1983). PROSHANSKY et al. assume "[...] that the development of self-identity is not restricted to making distinctions between oneself and significant others, but extends with no less importance to objects and things, and the very spaces and places in which they are found" (1983, p. 57). According to them, places define and give structure to the day-to-day life of individuals and thus shape the subjective sense of the self (PROSHANSKY et al., 1983, p. 58). In the development of a self-image, for example, personally important events and communities (work colleagues, friendship, family and kindship, etc.) connected to particular places play an important role (CONRADSON and McKAY, 2007, p. 168). Place identity is considered as a sub-structure of self-identity, which consists of cognitions (memories, feelings, meanings, behaviors etc.) about the everyday physical environment of an individual (PROSHANSKY et al., 1983, p. 59). Thus, place identity implies a process of self-reflection. This means that the self reflects its spatial involvement and analyses the personal meaning of the individual's environment (LENGEN, 2019, p. 123). However, place identity does not represent a coherent sub-structure of self-identity, as it should be thought of as a 'potpourri' of cognitions, which evolves through conscious and unconscious selective engagement with one's environment. Thus, place-identity is a personal construction of the (spatial) self, which modifies over time (PROSHANSKY et al., 1983, pp. 60 ff.; LENGEN, 2019, pp. 124 f.).

Practices and the spatial self
As place-identity and thus self-identity evolve through the engagement of individuals with their physical and social environment, it can be inferred that *Heimat*—if understood as a sense of spatial self—develops by everyday (social) practices. Practices can be understood as actions based on habits and routines, which are embedded in the implicit knowledge and skills of time-dependent, shared and local conventions as well as in the context of interrelated actions (SCHULZ-SCHAEFFER, 2010, pp. 320 ff.; RECKWITZ, 2003, p. 292). Thus, practices should be considered as a bundle of actions (SCHULZ-SCHAEFFER, 2010, p. 322). As previously stated with reference to GREVERUS (1972) and MITZSCHERLICH (2019), *Heimat* constitutes a space of security and action, which is characterized by familiarity and certain regularities, such as certain routinized actions, rituals and habitual social interactions (AL-ALI and KOSER, 2002b, p. 7). Such regularities or routines are required to convert place into *Heimat* (TUAN, 2012, p. 229). The concept of *Heimat*, thus, resembles the idea of an identity-forming bundle of actions, which is embedded in a space of knowing and identification. Thereby, *Heimat* itself is often considered as a practice, as it requires active engagement of individuals to build a space of positive identification. The place of residence, for example, can be considered as a cultural expression of different (social) practices, such as furnishing a home and 'curating' everyday life, which contribute to a sense of *Heimat* (EGGER, 2020, 31). In the case of changing the place of residence every action of an individual, which contributes over time to the adaption to one's physical and social environment and fosters the sense of belonging, can be summarized under the practice of *Heimat*(-making) (BINDER, 2020, p. 85).

Such practices require the individuals to draw similarities between the self and the physical as well as the social features of places, in order to develop a sense of

Heimat and to positively identify with places. In doing so, similarity and familiarity engender a sense of (local) belonging (SCANNELL and GIFFORD, 2010, p. 3; c.f. INALHAN and FINCH, 2004, p. 123). When it comes to belonging, TUAN, however, draws distinctions between a sense of place and rootedness. He argues that sense of place "[...] implies a certain distance between self and place which allows the self to appreciate a place" (1980, p. 6, as cited in EASTHOPE, 2004, p. 130). This means that the incorporation of certain qualities of a place into the self and the sense of belonging to a particular place "[...] is the result of conscious effort" (TUAN, 1980, p. 8, as cited in EASTHOPE, 2004, p. 130). TUAN writes that sense of place is a subjective measure of belonging because it is related to people's affinity to place (2012, p. 229). Rootedness, by contrast, implies a sense of belonging in an unselfconscious and unreflexive way. It is "[...] a knowing that is the result of familiarity through long residence" (TUAN, 1980, p. 8, as cited in EASTHOPE, 2004, p. 130). TUAN reasons, that rootedness can be considered as an objective measure, as "[...] archaeological and historical research shows that some people have lived in the same place for centuries" (2012, p. 229). Without making a distinction between rootedness and sense of place, the geographer John TOMANEY writes about the sense of belonging as follows:

> "Belonging concerns, simultaneously, feeling 'at home' and 'feeling safe'. A sense of local belonging can be expressed individually or collectively. Belonging can be attached to narratives of identity, but it may reflect also practical commitments, investments and yearnings. Moreover, the construction of place identities and terrains of belonging are juxtaposed; expressions of local belonging may embody a performative dimension which links individual and collective behaviour and contributes to the formation of narratives of identity and the realization of attachments. But belonging is formed in an intersectional context, along multiple, mutually constitutive axes of difference, of which geography is only one [...]" (2014, p. 508).

In contrast, Kathleen MEE and Sarah WRIGHT emphasize on the geographical dimension of belonging by writing in their guest editorial of a theme issue on 'Geographies of Belonging', that the concept of belonging is inherently geographical, as it "[...] connects matter to place, through various practices of boundary making and inhabitation which signal that a *particular* collection of objects, animals, plants, germs, people, practices, performances, or ideas is meant 'to be' in a place [...]" (2009, p. 772). Similar to the concept of *Heimat*, a sense of belonging refers to different spatial scales—the local, national or transnational (MORLEY, 2001, p. 425). The potential spaces of belonging, such as home, neighborhood, town or nation, however, should not be viewed as separate, but rather as mutually dependent (MORLEY, 2001, p. 433).

Belonging can be considered as the primary function of spatial identity (c.f. PROSHANSKY et al., 1983, p. 61) and thus constitutes an integral part of the sense of *Heimat*. Seen from this angle, *Heimat* can be understood as a practice of self-location. This also aligns with the considerations of JOISTEN, who is convinced that *Heimat* is part of human being, as people need to form (spatial) bonds in order to orient themselves. In her view, *Heimat* is an expression and convergence of fundamental phenomena of human existence: to locate oneself in space [*Sich-Orten*], in time [*Sich-Zeitigen*], and to encounter each other [*Sich-Begegnen*] (2003, p. 24). *Heimat* as a practice of self-location may imply that individuals define their 'place to be' through various (inter)actions based on their sense of belonging. A sense of belonging is often also connected to being committed to meaningful places and the people located in these places, since people tend to feel more responsible towards local issues (JOISTEN, 2003, p 26). Hence, *Heimat* as a practice of self-location always stands in an opposing relation to *Fremde*. However, the foreign, unfamiliar and unknown is not always located outside of *Heimat*, as encounters with *Fremde* also take place in *Heimat* (c.f. MORLEY, 2001, p. 428). This produces an "[…] interplay between safety and unsafety, security and insecurity, proximity and distance, trust and distrust" (JOISTEN, trans., 2003, p. 27). Following JOISTEN, *Heimat* constitutes a complex and dynamic entity, which is growing with its differences, tensions and contradictions and thus reveals its potentials (2003, p. 27).

2.2   *Heimat* in a Globalized World

In the face of globalization and processes of societal transformation, *Heimat* as a socio-spatial reference point seems to gain in significance (SEIFERT, 2010b, p. 21), although it is often assumed that person-place bonds become more fragile (SCANNELL and GIFFORD, 2010, p. 1). As already mentioned, *Heimat* offers not only the satisfaction of fundamental human needs such as security, safety and familiarity, but also offers orientation and simplification in an increasingly globalized, fragmented and complex world (KÖSTLIN, 2010, p. 36). In this sense, *Heimat* is often considered as an anthropological constant (RUNIA, 2020, p. 167) and functions as a metaphorical compass in the postmodern age (WEBER et al., 2019, p. 11). Usually, *Heimat* in German-speaking contexts is perceived as fixed depending on the 'place of origin', which refers to the place of birth and the place of childhood or the place where people spent most of their lifetimes. The question about *Heimat* is therefore usually discussed at the level of one specific territory, such as a town, a region or even a nation-state. Such territorial spaces are constructed according to the logic of positioning and arrangement, with clear boundaries outwards and ideas of homogeneity inwards. Due to such demarcations *Heimat* is often perceived as static (LÖW and KNOBLAUCH, 2021, p. 36). However, a static and monolithic understanding of *Heimat* needs to be reconsidered or even discarded, since it does not match the reality of many people living in a world that is shaped by globalization and migration flows (ARNOLD, 2016, p. 162; ECKER, 2012, p. 208). With the increase in national as well as international mobility the familiar *Heimat* changes and also intermingles with the *Heimat* of other people. As a consequence, the individual sense of *Heimat* can become irritated but also enriched, expanded, multiplied and relocated (WEBER et al., 2019, p. 12). Or, to put it in the words of anthropologist Nadje AL-ALI and geographer Khalid KOSER: "[...] socially homogeneous, communal, peaceful, safe and secure homes (Rapport and Dawson, 1998a) belong to the past (whether imagined or real)" (2002b, p. 7).

In the following chapters, the concept of *Heimat* will be contextualized within a globalized world. For this purpose, it will first be discussed how globalization processes, especially increased mobility and flows of migration, create a field of tensions between stability and change and refigurate locally conceptualized spaces such as *Heimat* (Section 2.2.1). Subsequently, a singular and static understanding of *Heimat* will be reconsidered in the light of international migration and translocal practices, which lead to a plurilocal sense of *Heimat* (Section 2.2.2).

2.2.1 *Heimat* in the Field of Tension between Stability and Change

In general, it can be stated that the concept of globalization describes a variety of transformative processes, as a consequence of human actions, which result in global networks and interdependencies (DÜRRSCHMIDT, 2004, p. 12; LÖW et al., 2021b, p. 10). These transformative processes include social and economic changes as well as increasing rates of migration flows (EASTHOPE, 2009, p. 65). The sociologist Anthony GIDDENS defines globalization "[...] as the intensification of worldwide social relations which link distant localities in such a way that local happenings are shaped by events occurring many miles away and vice versa" (1991b, p. 64). Globalization, thus, alters the relations of space and time, as increased flows of ideas, information, commodities and people transform the local and laterally extend social connections due to global translocal dialectical processes (GIDDENS, 1991b, p. 64). The world, thus, can be described as a 'space of flows' (FAIST, 2010, p. 1674). Over the past decades, social structures, spatial imaginations as well as daily spatial practices have been fundamentally reshaped (LÖW et al., 2021b, p. 10). This also applies to constitutions of identity (BLICKLE, 2002, p. 29; ROSE, 1995, p. 116). The alteration in the process of identity construction can "[...] be understood as a shift from relatively stable identities rooted in place to hybrid identities characterized by mobility and flux" (EASTHOPE, 2009, p. 65). According to the anthropologist Arjun APPADURAI (1996), globalization and migration flows lead to the dissolution of the close connection between spaces and group identities. In other terms, he assumes that globalization leads to deterritorialization (ECKER, 2021, p. 210). In order to examine the dynamics of deterritorialization, he explores in his work *Modernity at Large* "[...] the relationship among five dimensions of global cultural flows that can be termed (a) *ethnoscapes*, (b) *mediascapes*, (c) *technoscapes*, (d) *financescapes*, and (e) *ideoscapes*" (APPADURAI, 2005, p. 33). The suffix '-scape' indicates, according to him, that these landscapes are of irregular and fluid shape and are constructed from various perspectives (historical, linguistic, and political) of different actors, such as "[...] nation-states, multinationals, diasporic communities, as well as subnational groupings and movements [...], and even intimate face-to-face groups, such as villages, neighborhoods, and families" (APPADURAI, 2005, p. 33). APPADURAI further writes that the current global flows occur in and due to growing disjunctures among the mentioned landscapes. He also notes, that, of course, disjunctures in flows occurred at all periods in history, but the speed, scale and volume of flows has increased and as a result, the disjunctures have become a determining factor in the politics of global culture (APPADURAI, 2005, p. 37). In the context of *Heimat* the dimension of ethnoscapes is of particular interest (c.f. ECKER, 2012, p. 210), as it refers to

> "[...] the landscape of persons who constitute the shifting world in which we live: tourists, immigrants, refugees, exiles, guest workers, and other moving groups and individuals constitute an essential feature of the world and appears to affect the politics of (and between) nations to a hitherto unprecedented degree" (APPADURAI, 2005, p. 33).

APPADURAI further clarifies that

> "[t]his is not to say that there are no relatively stable communities and networks of kinship, friendship, work, and leisure, as well as of birth, residence, and other filial forms. But it is to say that the warp of these stabilities is everywhere shot through with the woof of human motion, as more persons and groups deal with the realities of having to move or the fantasies of wanting to move" (2005, pp. 33 f.).

Along a similar vein, the sociologist John URRY argues that contemporary forms of belonging "[...] almost always involve diverse forms of mobility" (2000, p. 132). According to him, "[...] people dwell in and through being both at home and away, through the dialectic of roots and routes [...]" (URRY, 2000, pp. 132 f.). Following his manifesto—which is centered on a sociology of flows instead of a sociology of territories—identities are mobile, diasporic and transient (SAVAGE et al., 2005, p. 1). Different theorists conclude that identities in a mobile world can be considered as dynamic (RUTHERFORD, 1990) and hybrid (BHABHA, 1994). Thus, in a mobile world, which is characterized by increasing migratory movements, shifting practices and cognitions, the meaning that is attached to the idea of nation-states, citizenship, and boundaries, which circumscribe the distributions of people and groups, is also changing (FAIST, 2012, p. 1; FAIST, 2010, p. 1675).

Refiguration of space: simultaneity of opposing trends
Increasing migration flows, nonetheless, do not necessarily lead to deterritorialization and the dissolution of person-place bonds. Migration rather leads to an alteration of the articulations of the relationship between place, identity, and thus of defining belonging (GILMARTIN, 2008, p. 1838). The sociologist and human geographer Hazel EASTHOPE argues that alterations in the process of identity construction do not imply that people no longer have some sort of spatial attachments and create meaningful places. Places still have an impact on the identities of people (2009, p. 66; c.f. DUYVENDAK, 2011). She underlines that both stability and change is important for the constitution of identity. On the one hand, place with its social, physical and cultural features contributes to maintaining a coherent identity. And on the other hand, mobility affects the development of identity (EASTHOPE, 2009, p. 77). At the same time, however, she also warns

"[...] against simply equating place with stability and mobility with change. Places, understood as nodes in networks of relations, are not stable in the sense of being static. Rather, they are constantly re-negotiated and understood in new ways by different people, or by the same people at different times. [...] Similarly, mobility need not necessarily imply change. For some people, mobility itself has become normalized; [...] It is important to recognize that reality is even more complex than these dialectics imply" (Easthope, 2009, p. 77).

The sociologist Martina LÖW and her colleagues also criticize in their anthology *Am Ende der Globalisierung: Über die Refiguration von Räumen* [At the End of Globalization: On the Refiguration of Spaces] (LÖW et al., 2021a), that such dialects are often approached naively. They refer in particular to theories on globalization, which tend to contrast the global with the local. Löw et al. argue that processes of globalization and localization, nationalization and internationalization, heterogenization and homogenization, the formation of networks and territorial closure and so on, should be considered as co-present phenomena, which stand in a tense relationship within a spatially changing modern society (Löw and Knoblauch, 2021, p. 26; Löw et al., 2021b, p. 10). Since the authors consider the concept of globalization insufficient to describe the polarizations, tensions, contradictions, and simultaneities of opposing trends, they establish the concept of 'refiguration' (Löw et al., 2021b, p. 10). The authors, however, do not intend to use the term as a replacement for the concept of globalization. Refiguration can be understood as a part of related concepts, such as 'glocalization' (Robertson, 1995), 'cosmopolitanization' (Beck, 2002; Beck and Sznaider, 2006), 'multiple modernities' (Eisenstadt, 1999) and 'entangled modernities' (Randeria, 1999; Therborn, 2003), which are used to describe the phenomena of globalization (Löw and Knoblauch, 2021, p. 25). The concept of refiguration thereby rather focuses on societal modifications triggered by tensions from a spatial perspective and emphasizes on the refiguration of spaces as a process on different levels (Löw and Knoblauch, 2021, p. 32).

Appadurai (2005) identifies the tension between homogenization and heterogenization as the central problem of global interactions. He writes, for example, that polities of smaller scale always fear to be culturally absorbed by polities of larger scale (Appadurai, 2005, p. 32). This tension and anxiety of absorption could also be transferred to the relation between *Heimat* and *Fremde*, as *Heimat* constitutes a personal socio-spatial reference point of small scale, whether it is referring to a physical place or a symbolic space (Al-Ali and Koser, 2002b, p. 7); whereas large-scale processes of globalization seem to endanger this personal unique space of security, safety and familiarity (c.f. Blickle, 2002, p. 14). Following this, a sense of *Heimat* usually concerns the level of the local. However, the essence and scale of the local often remains uncertain (Savage et al., 2005, p. 4). Appadurai argues that the boundaries of localities are defined by contexts. Thus, localities are socially produced by processes of boundary defining (2005, pp. 182 f.; Savage et al., 2005, p. 7). *Heimat*, for instance, involves contexts of everyday life routines, biographical references and identity as well as belongingness based on similarities.

Similar considerations can be found in the work of the cultural scientist David Morley, who writes that contemporary forms of mobility contest and "[...] transgress the boundaries of the sacred spaces of the home or *Heimat* [...]" (2001, p. 432; c.f. Glick Schiller et al., 1995, p. 50). The issues is, according to him, that the processes of how those transgressions are regulated generate conflicts, as it is usually attempted to exclude alterity. This raises the question of who defines alterity and thus who belongs to the 'sacred space' or what matters are considered to be 'out of place'. This could be, for example, 'strangers' or 'foreign' cultural objects and practices, which are perceived to adulterate the symbolic space of the *Heimat*—whether at the level of the home, the neighborhood, the region or the nation-state (MORLEY, 2001, p. 432). The fact that *Heimat* is produced through social, collective but also very personal processes of identity building and a sense of belonging underlines its subjective and rather irrational quality. More objective indicators of belongingness like forms of legal membership such as citizenship, which signals social closure and (usually) ensures the equality and certain (participative) rights of all members (FAIST, 2010, p. 1667), seem to play a less significant role. But citizenship is also not to be excluded as a factor for the sense of *Heimat*, as it is fundamental for inclusion and exclusion in terms of rights and obligations and thus participation, recognition and identification (FAIST, 2010, pp. 1679 f.; c.f. GILMARTIN, 2008, p. 2843).

The concept of refiguration allows for the examination of the processes of the formation, transformation and dissolution of global spaces as they are considered as historically constructed, multiple and overlapping spatial orders. In contrast to deterritorializing globalization theories, the concept of refiguration also pays attention to spatial orders on various spatial scales. For instance, smaller spatial scales such as national territories, regions and places receive attention for their role in the processes of globalization. On the one hand, the national, regional and local refigurate under the influence of globalization processes, but on the other hand, they also have an effect on such processes (LÖW et al., 2021b, pp. 13 f.). Thus, processes of globalization do not only involve "[...] the reconstruction of, in a sense the production, 'home', 'community' and 'locality'" (ROBERTSON, 1995, p. 30). The findings by LÖW and KNOBLAUCH (2021) suggest, in opposition to common assumptions, that places gain in relevance; not only by the bundling of heterogenous processes— the 'throwntogetherness' like the geographer Doreen MASSEY (2005, p. 140) would call it—but also through individual processes of place-making. Such individual processes can be found, for instance, in practices of self-location and home-making (LÖW and KNOBLAUCH, 2021, p. 38), and thus in the concept of *Heimat*.

2.2.2 A Plurilocal Sense of *Heimat* in the Context of International Migration and Translocal Practices

Mobility and migration are often considered to oppose settledness, as is apparent from the tension between change and stability. This dichotomous system of ordering is linked with the idea of settledness as the norm (LEHNERT and

LEMBERGER, 2015, p. 91). In fact, many people in the world never migrate to another country and stay in their home locality (GLICK SCHILLER, 2007, p. 458). The United Nations (UN) define people who spend at least one calendar year outside their country of birth or country of citizenship as international migrants (United Nations, 2020a, p. 5). According to the UN report on International Migration, there are 281 million people, who fall into the category of international migrants in 2020. These people constitute approximately 3.6 percent of the world's population (United Nations, 2020b). Despite this relatively small proportion of the world's population, it can be assumed that migration has been the norm over the course of human history (GLICK SCHILLER, 2007, p. 450). Also because the numbers of international migration have been steadily increasing over the past two decades (United Nations, 2020a, p. 1). This could be due to the reduction of the financial costs of migration (e.g. transport, accommodation, insurance, etc.) as well as international conventions that encourage the (im)migration of skilled and professional labor migrants (WICKRAMASINGHE and WIMALARATANA, 2016, p. 14). However, a clear determination of people's location in the sense of continuity and concordance between citizenship and place of residence tends to be viewed positively; while migration, in particular that of international migrants, is frequently stylized as a problem, for instance in discourses about integration/assimilation and the perceived threat of 'otherness' (LEHNERT and LEMBERGER, 2015, p. 91; YILDIZ and HILL, 2014b, p. 10; GLICK SCHILLER, 2010, p. 109; EHRKAMP and LEITNER, 2006, p. 1592). Moreover, mobility tends to be equated with rootlessness in this dichotomous system, whereas rootedness to a place is seen as belonging. Thus, mobility tends to be viewed negatively due to an assumed lack of belonging. In the German-speaking world, for example, migration, especially that of the so-called guest workers, was understood as a provisional condition and thus migrants were forced into a 'special role' (HILL, 2014, p. 174). However, this dichotomy of migration and sedentarism has been increasingly questioned by migration researchers in recent decades (BINDER, 2010, pp. 198 f.).

MORLEY argues that "[…] the question is not whether mobility or sedentarism are good or bad things in themselves, but rather of the relative power which different people have over the conditions of their lives" (2001, p. 430; c.f. MASSEY, 1995). In that sense, "[…] mobility is considered a fundamental aspect of social life and migration is considered a complex (and turbulent) process requiring a consideration on both structural factors and human agency" (EASTHOPE, 2009, p. 62). At this point, it becomes apparent that the creation of places, and thus the development of a sense of *Heimat*, is nonetheless not entirely subjective. It is also crucial to acknowledge the influence of the physical, economic and social realities of people (EASTHOPE, 2009, p. 70). Thus, places can be understood as "[…] doubly constructed: most are built or in some way physically carved out. They are also interpreted, narrated, perceived, felt, understood and imagined" (GIERYN, 2000, p. 465, cited after EASTHOPE, 2009, pp, 70 f.). One of the major differentiating factors regarding mobility is class, as both immobility as well as being forced into mobility, but also being able to be mobile (e.g. moving or travelling), are often related to economic factors (MORLEY, 2001, p. 429). This is also reflected in the UN Report on International Migration. In 2020, nearly 63 percent of all international migrants, and thus the majority, originate from middle-income countries (United Nations, 2020a, p. 1). The number of international migrants from low-income countries is rather small in comparison to other income groups. However, there has also been an increase in recent decades due to humanitarian crises, which is why the majority of international migrants from low-income countries are refugees and asylum seekers (United Nations, 2020a, p. 1). Forced migratory movement in consequence of economic predicament, such as poverty, lack of opportunities, or crisis situation, but also due to other reasons such as of political (e.g. persecution, conflict or war), social (e.g. family reunion) or environmental (e.g. drought or famine) nature, are commonly distinguished from voluntary forms of mobility (MORLEY, 2001, p. 430; KOSER, 2007, pp. 16 f.). The reason why people leave certain places and move to another place within a region, country or to another nation-state, can also be framed by push and pull factors. Both mainly emphasize on economic determinants. Push factors include the above-mentioned economic reasons for involuntary migration, while pull factors can include job opportunities and the prospect of improving the standard of living (GILMARTIN, 2008, pp. 1838 f.). In 2020, the region of Europe received 87 million people, and thus the largest number of international migrants. However, 70 percent of them are intra-regional migrants because they were born in one European country and live in another. North America received a total of nearly 59 million migrants. This is followed by Northern Africa and Western Asia with nearly 50 million international migrants in total (United Nations, 2020a, pp. 1 f.). However, according to the UN, Northern Africa and Western Asia could overtake North America within the next decades, if the current trend continues. In this regard, the UN writes that "[t]his shift reflects the increasing diversification of economic opportunities available to migrant workers and it foretells the greater competition that destination countries will likely face in the future to attract migrants, especially skilled migrants" (United Nations, 2020a, p. 1).

The differentiation between voluntariness and involuntariness is, nevertheless, not always clear, but is often more blurry in reality. Migration cannot be reduced to economic push and pull factors only and much depends on perspective. For instance, subjective experiences based on the legal status of migrants (KOSER, 2007, p. 17) as well as the gendered, sexual, racial or ethnical identity of individual migrants also give reason to mobility and shape the sense of *Heimat* and belonging (GILMARTIN, 2008, pp. 1839 f.). Thus, categorizations of migration forms as well as a clear distinction of migrant types tend to oversimplify reality and lack of a holistic approach (KOSER, 2007, p. 18; WICKRAMASINGHE and WIMALARATANA, 2016, p. 27). LEHNERT and LEMBERGER thus regard both poles—migration and sedentariness—as "[…] just two hypothetical endings of a continuum of daily practices and experiences, which likewise can transcend state borders and can be situated locally" (LEHNERT and LEMBERGER, 2015, pp. 91 f.). On the one hand, they plead for giving more visibility and attention to unmarked positions such as settledness, while on the other hand emphasizing that migrant experiences should be perceived as a part of a national narrative (LEHNER and LEMBERGER, 2015, pp. 94 f.).

Simultaneous embeddedness
The dichotomous juxtaposition of migration and sedentariness also simplifies the relationship between the 'here' and the 'there', although this is becoming increasingly unclear in the course of international migration (AL-ALI and KOSER, 2002b, p. 8). In other words, this juxtaposition ignores simultaneities, such as living across national borders and being simultaneously embedded into two or more nation-states. International migrants' practices of self-location and place-making are rarely monolithic and spatially singular, but rather multilayered and plurilocal as transnational perspectives on international migration suggest (c.f. BASCH et al., 1994; GLICK SCHILLER et al., 1992, 1995; GLICK SCHILLER, 2007, 2010; FAIST, 2010, 2012). The paradigm of transnationalism does not form a coherent set of theories (FAIST, 2012, p. 1), but it rather offers a 'conceptual space' (GLICK SCHILLER, 2007, p. 448) for different disciplinary approaches (AL-ALI and KOSER, 2002b, p. 2) to encompass the experiences and life realities of so-called 'transmigrants', which can serve here as an example of the fact that migration movements cannot always be clearly divided into involuntary and voluntary forms of mobility and that the construction of a sense of *Heimat* (including the sense of belonging and identity) is subject to complex processes and practices of multiple self-location.

The term transmigrants was introduced by the anthropologist Nina GLICK SCHILLER and her colleagues (GLICK SCHILLER et al., 1992; BASCH et al., 1994) and refers to "[…] immigrants [and their descendants] whose daily lives depend on multiple and constant interconnections across international borders and whose public identities are configured in relationship to more than one nation-state" (GLICK

SCHILLER et al., 1995, p. 48). In other words, transmigrants live 'in between' at least two nations (KOSER, 2007, p. 27). This means that these people are settled and well incorporated into the economic and political institutions as well as the patterns of everyday life of the country where they live. But, simultaneously, these migrants are agents within the constellations of cross-border flows as they also maintain connections and get involved in a participatory way in local and national affairs of the countries from which they emigrated, be it on a political, economic or social level (GLICK SCHILLER et al., 1995, p. 48.; FAIST, 2012, p. 3; FAIST, 2010, p. 1673). A complete assimilation, in the sense of breaking off all social relations and cultural connections to the country, which was left, hence does not take place (FAIST, 2012, p. 2). Thus, transmigrants are not rooted in a single place, but they are embedded in more than one society (GLICK SCHILLER et al., 1995, p. 48) through "[...] the ongoing interconnection and flow of people, ideas, objects, and capital across the borders of nation-states [...]" (GLICK SCHILLER, 2007, p. 449).[1] However, it must also be said that it is not always clear which international migrant can be called a transmigrant. Not all international migrants automatically live a transnational life. Most migrants are only occasionally transnational actors (LEVITT, 2004). According to the UN, members of diasporic communities, nevertheless, play an important role and contribute decisively to the development of the countries they have left, for example, by supporting foreign investments, trade, innovations and so on (United Nations, 2020a, p. 1). This simultaneous process of shaping, sustaining and linking

---

[1] With regard to the study of transnational processes, GLICK SCHILLER warns against, what she calls, methodological nationalism. According to her, this intellectual orientation "[...] assumes national borders to be the natural unit of study, equates society with the nation-state, and conflates national interests with the purpose of social science" (2007, p. 451; c.f. WIMMER and GLICK SCHILLER, 2002). Nation-states and borders should always be understood as social, political and historical constructs as they are "[...] constructed within a range of activities that strive to control and regulate territory, discipline subjects, and socialize citizens, but these processes and activities are not necessarily located within a single national territory" (GLICK SCHILLER, 2007, p. 449). The fallacy is, on the one hand, to assume that all members of a national state share certain commonalities, such as a common history and a set of social norms and values (GLICK SCHILLER, 2010, p. 111). A nation can therefore also be described as an 'imagined community' (ANDERSON, 2006). And on the other hand, transnational approaches should overcome defining migrants in terms of categories such as ethnicity and nation, as migrant formations can also be built around other distinctive categories, such as class, race, gender, sexual orientation, (dis)ability, political affiliation, educational status and so on (FAIST, 2012, p. 4). Sociologist Thomas FAIST, however, emphasizes that the national should not be understood as something that is to be overcome as the nation state is still a site of the assertion of rights. He decisively rejects the idea of a decreasing importance of nation states and points out that transnationalism should not be confused with post-nationalism (2010, p. 1672).

the multi-stranded social relations in both the societies of origin and settlement is referred to as transnational migration (GLICK SCHILLER et al., 1995, p. 48). According to GLICK SCHILLER, the nation-state does not contain but shapes interconnections and flows across borders (2007, p. 449). Such transnational processes are rather located in between "[…] the functional systems of differentiated spheres, such as the economy, polity, law, science and religion" (FAIST, 2010, p. 1673). In particular, it is the interpersonal social relationships and institutional linkages that form cross-border networks through the exchange of information, resources, ideas, and so on. These networks of cross-border interpersonal connections do not only take place within the framework of the nation-state, but rather within transnational social fields (GLICK SCHILLER, 2007, p. 457; GLICK SCHILLER, 2010, p. 112). These transnational social fields contain, for example, institutions, organizations, but also experiences which generate different categories of identity, not automatically based on nationality (LEVITT, 2004). The UN indicate that the financial support of transnational migrants in the form of investments, for example, in the areas of education, health or infrastructure, contributes decisively to improving the livelihoods of families and communities in their countries of origin (United Nations, 2020a, p. 1; c.f. LEVITT, 2004). But these processes of transnational migration are also situated in between the experiences, relationships, and personal interactions of individuals, families, kinship and other groups (GLICK SCHILLER et al. 1995, p. 50). These social agents constitute the smallest level of transnational social formations (FAIST, 2012, p. 1). The maintenance of cross-border social relationships is made possible and facilitated in particular by modern digital information and communication technologies (HILL, 2014, p. 174). However, transnational life is not only performed through people-to-people relationships, but also through the means of communication: "In reading a book, newspaper, or magazine, listening to a radio, watching a film or television, or surfing the internet one can obtain ideas, images, and information that cross borders" (GLICK SCHILLER, 2007, p. 457).

If one now wants to take a closer look at the motives for their migration in order to assess how much power these transmigrants have over their life circumstances, one finds different perspectives in literature. On the one hand, scholars refer to labor migrants who migrate from economically less developed countries to countries that offer greater economic security. In the same breath, (political) refugees fleeing conflicts and instability are also often mentioned (c.f. FAIST, 2012, p. 1). GLICK SCHILLER et al. identify three conjoining potent forces, which lead to transnational migration. According to them, one reason for a transnational life is the global restructuring of capital, which has resulted in the deterioration of social and economic conditions in labor sending as well as labor receiving countries. As a result, no country is considered to offer a secure place to settle (1995, p. 50). Another force

GLICK SCHILLER et al. identify is racism. Due to racial discrimination in receiving countries, immigrants and their descendants might experience economic and political insecurities. The third force favoring transnational life is found in nation-building projects, which contribute to the immigrants' political loyalties to each nation-state (GLICK SCHILLER et al., 1995, p. 50; c.f. LEVITT, 2005). But on the other hand, studies on transnational migration, or transnationalism in general, have also often referred to elite transmigrants and their power. Transnational elites are characterized by the fact that they have access to capital and thus are able to bypass restrictive legislations on immigration and to make policy decisions (GILMARTIN, 2008, p. 1841; c.f. MITCHELL, 2003; KAUPPI and MADSEN, 2013).

The motives for migration, whether they are based on political, economic, social, or environmental reasons, can be important to the development (or absence) of a transnational identity and the extent to which an international migrant engages in transnational activities. Factors such as class, race, gender, sexuality, age, (dis)ability, political affiliation, religion, educational status and so on, of course also play a very large role in this process (AL-ALI and KOSER, 2002b, p. 3). The level of involvement in the country of origin also depends on the stage in one's life (LEVITT, 2004). Thus, the practices, identities and sense of belonging of transmigrants are not homogeneous (EHRKAMP and LEITNER, 2006, p. 1594). However, regardless of how much power these migrants have over the conditions of their lives and why they cross borders, their life reality becomes a transnational one through different transnational practices, as already described above. Here, GLICK SCHILLER emphasizes that it is essential to distinguish between "transnational ways of being" and "transnational ways of belonging" (2007, p. 458). By transnational ways of being, she refers to various everyday actions by which people live their lives beyond nation-state borders. This does not necessarily mean that the members of a transnational social field have to physically cross borders themselves, but in any case it is the interactions that take place across borders. Thus, a transnational way of being refers to a way of life (GLICK SCHILLER, 2007, p. 458). This transnational way of life bridges and transcends legal and political geographies as well as social relationships and cultural meanings, norms, values and practices (BLUNT, 2007, p. 688). However, this way of life does not indicate how the practices of transmigrants are represented, understood, and translated into identity and a sense of belonging (GLICK SCHILLER, 2007, p. 458). Therefore, she separately refers to transnational ways of belonging in order to explore "[...] the realm of cultural representation, ideology, and identity through which people reach out to distant lands or persons through memory, nostalgia, and imagination" (GLICK SCHILLER, 2007, p. 458). Further she explains that

> "[t]ransnational belonging, while not rooted in social networks, is more than an asser-
> tion of origins, optional ethnicity, multiculturalism, or "roots," which are all forms of
> identity which place a person as a member of a single nation-state. Ways of belong-
> ing denote processes rather than fixed categories. Persons who adopt certain forms
> of cultural representation may find themselves as new participants in transnational
> social fields. [...] On the other hand, persons who live in transnational social fields
> may adopt, at various times, different forms of cultural representation. Transnational
> belonging is an emotional connection to persons who are elsewhere – a specific local-
> ity such as a village, a region, a specific religious formation, a social movement – or
> are geographically dispersed but bound together within a notion of shared history and
> destiny" (GLICK SCHILLER, 2007, p. 459).

Simultaneity is thus also reflected in the practices of transmigrants that signal or enact their identities and sense of belonging (LEVITT, 2004). Depending on their transnational embeddedness and associated practices, the identities of international migrants can be described as fluid and multiple (FAIST, 2012, p. 2). It seems that these migrants evade limiting political categorizations, such as citizen and immi-grant (KOSER, 2007, p. 27), and even establish their own identities based on different spatial ties they can draw on (ESCHER, 2006, p. 62). By the convergence of several different social and cultural practices in one place at one time, it can also some-times lead to the hybridization of several different social and cultural identities of an individual (ARNOLD, 2016, p. 165). By "[n]egotiating, improvising, experiment-ing and adapting, they take on contextually specific identities, emphasising and underplaying particular aspects of their being [...]" (SVAŠEK and DOMECKA, 2012, p. 107). But at the same time, as GLICK SCHILLER et al. also already note in an earlier publication, transnational activities seem to be accompanied by the revitalization, reconstruction, or reinvention of traditions but also of political claims in terms of territory and history. Thus, transnational migration also re-inscribes the migrant's identity onto the territory of the country that was left (1995, p. 52). Thus, processes of de-, re- and nationalization are constantly occurring simultaneously (HILL, 2014, p. 174).

Heimat as a practice of plural self-location

The relationships between the places of origin and the places of destination deter-mine routine practices, daily lives and self-conceptions of many migrants in such a way that it is difficult for many of them to clearly locate themselves in just one place (RALPH and STAEHELI, 2011, p. 519; CONRADSON and MCKAY, 2007, p. 168). Through refigurative processes of migration, the biographies as well as the mean-ingful places of the migrants have been set in motion sustainably (HILL, 2014, p. 173). Therefore, processes of transnational migration can also be referred to as 'translocality', with this term accentuating the spatial dimensions of multiple social and emotional embeddedness more strongly. Due to this, and in order to avoid methodological nationalism, in the following 'translocal' is preferred to the term 'transnational'. The notion of translocality was coined by APPADURAI (1996) and "[...] describe[s] the ways in which emplaced communities become extended, via the geographical mobility of their inhabitants, across particular sending and destination contexts" (CONRADSON and MCKAY, 2007, p. 168). In reference to APPADURAI (1996), the geographers David CONRADSON and Deirdre MCKAY employ the term 'translocal subjectivities' in order "[...] to describe the multiply-located senses of self amongst those who inhabit transnational social fields" (2007, p. 168). According to them, firstly, translocal subjectivities emerge through both mobility and multiple, simultaneous and ongoing forms of embeddedness in the different local contexts. This acknowledges persistence of emotional and social attachments towards particular places and the social relations within. Secondly, they note that the identity of international migrants is usually more closely related to specific people and places within a nation-state rather than to the abstract construction of a nation. And thirdly, they emphasize the importance of emotions and affects that accompany mobility (CONRADSON and MCKAY, 2007, pp. 168 f.).

The translocal geography of multiple and simultaneous embeddedness is, of course, also reflected in the people's conception of *Heimat*, which can be "[...] messy, mobile, blurred and confused" (RALPH and STAEHELI, 2011, p. 519). On the one hand, international movements of people, which also involve processes of establishing as sense of belonging and identity, create new places (BLUNT and DOWLING, 2006, p. 2). On the other hand, at the same time, through international migration, translocal practices and the simultaneous embeddedness in the 'here' and 'there' *Heimat* becomes a dynamic social space characterized by plural emotional and social attachments to places as well as a plurilocal sense of identity and belonging (AL-ALI and KOSER, 2002b, p. 6). The realities of life for many international migrants thus also challenge the concept of *Heimat* that is usually not only understood as fixed and bounded (RALPH and STAEHELI, 2011, p. 518) but is also used in the singular or is considered as unique, which is implied by speaking of a person having found a 'new' or 'second *Heimat*' (BLICKLE, 2002, p. 63). Through the plural self-location of international migrants, it is thus also possible to refer to the pluralized '*Heimaten*'. However, this plural form of *Heimat* is not used in the theoretical framework of the present study, as it implies that there is a clear distinction between the 'here' and the 'there'. There is one *Heimat* where one is right now, and there is at least one other *Heimat* somewhere else—one cannot (physically) be in both at the same time. Thus, *Heimaten* tends to exclude a simultaneous embeddedness in more than one place. Rather, *Heimat* is conceived in this study as a space that is expressed across borders through various everyday social practices and cognitive processes. It is "[…] a space, a community created within the changing links between 'here' and 'there'" (AL-ALI and KOSER, 2002b, p. 6). This means that the word in its singular form can be considered containing many different processes and plural spatial references that are linked to identity, belonging and certain experiences and emotions. Thereby however, the sense of *Heimat* does not remain free of 'contradictions', as migration and translocal practices imply a dynamic and ongoing process of '*Heimat*-making' and the negotiation of the contradictions of both places of embeddedness (STAEHELI and NAGEL, 2006, p. 1599). This dynamic process involves "[…] the acts of imagining, creating, unmaking, [maintaining], changing, losing and moving 'homes'" (AL-ALI and KOSER, 2002b, p. 6). Spatiality always seems to play a very large role in translocal identities, as implied by the negotiation between the processes of de-location and (re)location (c.f. GEORGIOU, 2006, p. 10).

How migrants become embedded in an unfamiliar place and thus transform it into a familiar place, cannot be expressed in general terms, as there are many different factors. The degree of local involvement of international migrants is, for example, decisively influenced by political forces that not only permeate spaces and places, for example in institutionalized settings, but also determine the possibilities for action and subject positions of migrants (BINDER, 2010, p. 202). Other essential aspects that determine the degree of embeddedness in local contexts and thus also contribute to the construction of *Heimat* are the social and emotional experiences but also personal life plans and conditions (BINDER, 2010, p. 202) as well as the participatory engagement in local and social matters resulting thereof (GILMARTIN, 2008, p. 1845). Especially emotional processes play an important role as they connect the individuals to human as well as non-human environments, and also shape social interactions and practices (SVAŠEK, 2010, p. 876). On the one hand, moving to an unfamiliar environment may evoke different feelings or even conflicting emotions, such as excitement, hope, joy but also fear or anger (SVAŠEK, 2010, p. 866). Among the typical experiences that international migrants have in a new environment are the confrontation with a language barrier and the general hardship of leaving a familiar place, reorienting themselves and establishing a daily routine or building a 'new life' (ARMSTRONG, 2004, p. 244). Other decisive emotional experiences are the encounters between the members of local communities and the newcomers, which are influenced by memories and expectations and thus may "[…] for example, stimulate or discourage migrant identifications with their new surroundings" (SVAŠEK, 2010, p. 867). And on the other hand, the emotional attachments of international migrants to the countries and places they left may evoke the need to sustain social relationships and a sense of family (SVAŠEK, 2010, pp. 866 f.). *Heimat*-making practices of migrants also include a variety of practices that 're-create' a familiar environment, such as the use and arrangement of familiar material objects (ECKER, 2012, p. 212), food preparation, consuming media or speaking a familiar language (SANDU, 2013, p. 509). The geographers David RALPH and Lynn A. STAEHELI write, that "[t]he construction of home [i.e. *Heimat*] is thus not necessarily tied to a fixed location, but emerges out of the regular, localising reiteration of social processes and sets of relationships with both humans and non-humans" (2011, p. 519). In this process, traditional practices can be preserved, others 'left behind', or even transformed (SANDU, 2013, p. 509). This process of *Heimat*-making is a lifelong practice of constantly locating and connecting oneself with places, people and other reference systems (MITZSCHERLICH, 2019, p. 188).

2.3 Operational Definition of *Heimat*

It is not possible to pose a universal and all-embracing definition of *Heimat*. The understanding of *Heimat* varies between different (inter)disciplinary approaches, but above all between the subjective perspectives of individuals, which are in the focus of this study. Some dimensions of *Heimat*, however, repeatedly appear in academic as well as public discourses. This allows for a narrowing of the phenomenon. The central aspects, which contribute to the development of a sense of *Heimat* can be linked to space, and thus can be examined from a geographical perspective. An operational definition, which serves as the framework for the development of the methodological approach and the analytical perspective on the gathered information, can be derived from the previous reflections on the concept of *Heimat* in the present study. This operational definition represents a synthesis of the considerations from an emotional-geographical perspective, the considerations on the interrelation of space, practices and self-conceptions, and the considerations on the refiguration of socio-spatial reference systems through (international) mobility.

Primarily, *Heimat* is an emotional space that is usually positively charged and conveys a feeling of safety, security and familiarity to individuals. These connotations are based on a positive identification with places, which is expressed in an emotional attachment to them and thus in an attribution of meaning. In addition to the physical characteristics of places, which are perceived by individuals with all their senses, in-place experiences as well as social connections and interactions play an important role in the formation of emotional people-place bonds. Places acquire meaning only through the cognitive processes of individuals, such as remembering, interpreting and imagining. The factor time is an important influence here, because on the one hand many formative experiences originate from childhood, on the other hand the duration of living contributes to the meaning of places, and furthermore the (re)interpretation of places and thus their meaning is only made possible by distancing oneself from them in space and time. However, *Heimat* is not only the product of cognitive processes, but is also constructed through various everyday routinized actions and habitual social interactions. These practices not only create a familiar environment but also always locate the individual in a particular socio-spatial context. This socio-spatial self-location as well as the social and emotional attachments as a result of cognitive processes contribute decisively to an individual's sense of self and sense of belonging.

The individual sense of *Heimat* thus represents the spatialization of an individual's biography, and thus *Heimat* can also be viewed as a socio-spatial reference system. This reference system is extremely rarely limited to a specific place, as it is oriented along the individual biography, which in turn is shaped by multiple places. Through mobility and being embedded in different local contexts, spatial references and thus the feeling of *Heimat* are constantly renegotiated. Social and emotional connections extend across multiple places, and also across national state borders. In this theoretical framework *Heimat* is thus not defined as a static, monolithic, and bounded socio-spatial reference system, but as a trans- and plurilocal and dynamic socio-spatial reference system.

Methodological Approach

3

This study will look at *Heimat* through the lens of international migration. The approach of considering migration not as a research object but as a perspective on other phenomena can be positioned within the concept of postmigration, which refers to postcolonial discourses (RÖMHILD, 2014, p. 6). In postmigrant discourses, migration is understood as a society-moving and society-shaping force. To approach phenomena from the perspective of migration means to visualize experiences of migration and thus marginalized ways of knowing. In addition to the goal of liberating conventional migration research from its previous 'special role' and establishing it as a social analysis, the postmigrant perspective also pursues the goal of retelling migration history. This retelling refers, on the one hand, to later generations of migrants, especially in relation to so-called guest workers in German-speaking countries. And on the other hand, the retelling of migration history refers to processes of transnationalization (YILDIZ, 2014, pp. 21 f.). This makes the postmigrant perspective linkable to transnational (GLICK SCHILLER et al., 1992; FAIST, 2012) as well as translocal perspectives (APPADURAI, 1996).

Selection of participants

Since the individual sense of *Heimat* is strongly linked to the biography of people, this study approaches the concept of *Heimat* through the sense- and meaning-making personal experiences, memories, stories, emotions, practices and subjective knowledge of international migrants (c.f. MAROTZKI, 2004, p. 103). For this study, the selection of participants focused on people who have moved at least once in their lives to another country for a long-term stay, which differs from the previous residence in the country of origin in various aspects, such as the predominant language spoken there and other social practices. Thus, people who move to another country for a short-term and planned time-limited stay, such as tourists, international students, and expatriates, are excluded as participants in

J. Andel, *Sense(s) of Heimat*, BestMasters,
https://doi.org/10.1007/978-3-658-38985-7_3

this study. Furthermore, refugees are also largely, but not completely, excluded from this study, as flight can be considered as a particular form of inter- and transnational migration (BINDER and TOŠIĆ, 2005). This exclusion is also justified, on the one hand, with regard to research ethical questions, such as dealing with trauma, and, on the other hand, by the additional complexity that the topic of flight in relation to the sense of *Heimat* brings with it, which would go beyond the scope of this study. Studies on *Heimat* and geographies of belonging in the context of flight experiences can be found, for example, in HUIZINGA and VAN Hoven (2018) and KÜCK (2021).

Another premise regarding the selection of participants is language. Since this study refers to the German term *Heimat*, the participants must be familiar with the German language to a certain degree. This also means that the study is to be placed in the German-speaking context. Thus, the focus is on international migrants who have either moved to a country in the German-speaking area or have moved away from such a country. The participants are mainly from the author's personal environment. However, there was only occasional to infrequent contact with most of the participants, so that only little was known about their personal biographies. In addition, some participants were completely unknown to the author beforehand, as they were recommended to her through common acquaintances. In total, there were twelve participants, some with very different and some with similar biographies. Some migration biographies are rather to be classified as voluntary migration, for example, because work-and-travel, educational opportunities, or romantic relationships that resulted in a long-term relocation to another country. Other migration biographies, however, are rather to be classified as involuntary migration, for example due to the family moving at a young age, for economic reasons, or for other personal reasons. However, the classification into involuntary and voluntary is not always unambiguous and is not made by all participants themselves. Besides Germany and Austria, the participants had a connection to Canada, Colombia, Greece, Lithuania, New Zealand, North Macedonia, Portugal, the United Arab Emirates (UAE), the United States (USA) and former Yugoslavia (today Slovenia and Croatia). Of the twelve participants, nine were female and three were male. The age ranged from 18 to 38 years, with the average age of the participants being 29 years. In addition to the age of the participants and other temporal factors, such as the age at the time of migration and the period of time a person has lived in a place so far, vary in some cases considerably. The twelve selected participants with their individual migration biographies can of course not cover all experiences of international migrants in relation to their sense of *Heimat*. They represent only a small extract of life realities as well as only a momentary insight into their own biographies.

Method and implementation of biographical online interviews

To obtain qualitative research data on the individual sense of *Heimat* of international migrants, the method of biographical interviews was used. Since the
biographies of the participants were very different and not everyone is used to
talk about their own biography, the interviews followed a combination of a semi-
structured and a narrative approach (as recommended in HOPF, 2004, p. 205),
whereby the participants were not confronted with standardized questions, but
were encouraged to speak openly about their experiences, thoughts and feelings.
Narratives are considered to be the most significant medium to express the self
(DE FINA, 2015, p. 351). The narrative biographical approach thus allows "[...] to
expose their individual worlds from the inside out. Consequently, the researcher
creates access to individual routines, practices, habits, and lifestyles of individuals [...]" (ARNOLD, 2016, p. 165). When conducting narrative interviews, it is
important that the main narrative comes from the interviewees themselves. The
interviewer should intervene as little as possible and only become more active
by asking internal and external questions towards the end (HOPF, 2004, p. 206).
However, it is not possible to conduct an interview that is free of any structure (MASON, 2002, p. 231). Relying only on the narrative approach turned out
to be inappropriate for most of the interviews, as most of the participants had
previously reflected little on their biographies and were not used to speaking at
length about themselves and their experiences. This is also reflected in the varying
length of the interviews. The shortest interview lasted 27 minutes, whereas the
longest interview lasted one hour and 46 minutes. On average, an interview lasted
47 minutes. Factors such as the familiarity between interviewee and interviewer,
the age of the participants, or their language level may also have played a role.
Although all participants understood and could speak German, three participants
felt uncomfortable conducting an interview in German, so these three interviews
were conducted in English.

In order to bring some structure to the interviews and thus to ensure a com-
parability of the data obtained, firstly, all participants were asked weeks before
the interview to bring something to the interview which they personally associated with the concept of *Heimat*. This 'bring along' could be something material
or immaterial and served as a conversation starter. And secondly, the interview
was partially guided by a mind map with relevant thematic fields, which was
developed deductively from theory and was inductively expanded by categories
in the course of the interviews. However, this guideline was only intended to
serve as a backup in case an interview faltered, and certain relevant topics were
not addressed by the participants. Thus, the interview was mainly intended to be
led by the participants, but there was also the possibility to flexibly adapt the

interview style to the interviewees. In all interviews, an attempt was made to create a pleasant atmosphere by having the interviews on a more casual and personal conversational level. It was made clear to the participants that they should only reveal what and how much they wanted to. In order to support the participants in narrating their biography, they were encouraged to talk about certain periods of time, for example by asking them to talk more about their childhood experiences, the migration process and the time of arriving and settling down. While some participants needed to be asked more questions on the basis of the guideline, others talked extensively about their experiences on their own initiative. The method of the ero-epic conversation [*ero-episches Gespräch*] developed by the sociologist and cultural anthropologist Roland GIRTLER (2001) served as an orientation for the interviews. In this method, the interviewees are encouraged to talk about themselves, but are not pressured to answer, as may be the case in classical narrative interviews. The aim is to consider the participants as equal conversation partners and experts in relation to the topic. This is also to be achieved by the researcher involving him-/herself, for example, by transparently talking about his/her research methods, research interests or own topic-related experiences. Suggestive questions as well as counter-questions from the interview partners are also allowed, as they can be equally revealing (HALBMAYER und SALAT, 2011). The communication before and after the actual interview can also reveal additional information and should be included in the later analysis (SVAŠEK and DOMECKA, 2012, p. 108).

For biographical interviews, it is common to ask the interviewees to choose a location where they feel comfortable (SVAŠEK and DOMECKA, 2012, p. 108). However, due to the current coronavirus pandemic, the interviews were moved to the digital space and were conducted via the platform BigBlueButton, which ensures a high level of data protection. In addition to the possibility to record and save the interviews through the platform, online interviews also offer the possibility to talk to participants who would otherwise not be available for a face-to-face interview due to the large distance of their current residences within Germany or in another country. In addition, all online interviews were conducted on a computer in an environment that was familiar to the participants, mostly in their own homes. Since qualitative research depends on trust, it is important to protect the privacy of the participants, to work in a way that is transparent to them, and to obtain their consent (SALMONS, 2016, pp. 62 f.). Therefore, prior to the interviews, all participants were informed about the procedure of the interview, that the interviews would be recorded for the purpose of transcription, their data would be kept confidential, and their names would be anonymized in the study by referring to them through an assignment of interviewee numbers (IP1-12).

After the interviews were completed, they were transcribed in detail using the transcription program F4.

Method and implementation of solicited online diaries
Another qualitative method to obtain qualitative research data in this study is the use of solicited online diaries, which can be seen as a complement to the biographical interviews. Solicited diaries are a common strategy to collect data within health related research, but rather uncommon in human geographic research (COHEN et al., 2006, p. 164; METH, 2003, p. 196). Unlike private diaries, which are questionable as a research source for ethical reasons, solicited diaries are not intended for private use, but are written with full awareness of an external observer. Of course, this offers the risk that the writers tend to write what the researcher wants to read. In general, however, solicited diaries offer many advantages. Firstly, such diaries generate a more empowering relationship between participant and researcher, because the participants are not given much information about the content and thus can contribute their own ideas that the researcher may not have thought of. This creates a fruitful collaboration, where new aspects in relation to the research topic can arise. Secondly, a diary offers participants another way to express their thoughts and experiences, as they have more time to reflect and formulate them in a written form (METH, 2003, pp. 196 f.). And thirdly, solicited diaries provide a valuable tool for gaining real-time insights into the emotions, thoughts, and impressions of the participants' current daily lives over an extended period of time, which are otherwise rather difficult to access (FILEP et al., 2015, p. 461; CLARK, 2020). Thus, solicited diaries are a useful complement to interviews, which offer only a relatively brief research encounter (SVAŠEK and DOMECKA, 2012, p. 110).

The diaries were kept online using simply designed Google Forms documents, which were deleted for privacy reasons after the data were analyzed. Each day over a two-week period, participants received a link for a new entry. In addition to a text box, participants also had the option to upload photographs. The participants were only instructed that the topic of the diary should be about *Heimat* in their everyday life. They were not asked to submit an entry every day, but to write at least seven entries. In addition, participants were free to choose whether to write the entries in English or German. This method was, however, not linked to the interviews in an obligatory way, since participation must be based on voluntariness and the participants must regularly take time to write their diary entries. Of all twelve participants, six people initially agreed to participate for the online diaries. Two of them, nevertheless, cancelled at short notice due to time reasons and in the end there were mainly two participants who wrote regular and detailed diary entries. A total of 22 diary entries of varying lengths were submitted. Therefore, the data from the diaries contribute only a very small part to this study.

Method of analysis

In order to analyze the qualitative data from the transcribed interviews and the diaries, the method of qualitative content analysis was employed. Through this method "[...] it is possible to distil words [from written, verbal or visual communication messages] into fewer content-related categories" (ELO and KYNGÄS, 2008, p. 108) and to attain 'a sense of whole' (BURNARD, 1991). This means that the goal of qualitative content analysis is to describe a phenomenon through the analysis of condensed concepts or categories (MAYRING, 2010, p. 62; ELO and KYNGÄS, 2008, p. 108). For this purpose, the research material should first be read intensively and repeatedly. Then, guided by the prior theoretical knowledge and the research question, all emerging themes and aspects relevant to the research question should be filtered, marked and condensed, in order to form categories and subcategories (c.f. SCHMIDT, 2004, pp. 254 f.). This part of the analysis may be described as 'formulating interpretation' (c.f. NOHL, 2010, p. 203). In this process, it is important to describe particularly prominent and frequent expressions as well as forms of theoretical value to make statements about the material and the formulated categories (MAYRING, 2010, p. 103). Similarities as well as differences between the interviews and diaries may also be relevant for further analysis (SCHMIDT, 2004, p. 254). The detailed elaboration of the emerged categories could be considered as 'reflecting interpretation' (c.f. NOHL, 2010, p. 203). First, the relevant text passages in the texts of the interviews and diaries must be assigned to the analytical categories and then described very distinctively. Subsequently, the research results are interpreted in detail (SCHMIDT, 2004, pp. 255 f.). Qualitative content analysis is more complex than formulaic and quantitative analysis methods and, thus, less standardized and very flexible. Therefore, there is no 'right' way to analyze the data. Rather, the results depend on the skills, insights, and style of the researcher (ELO and KYNGÄS, 2008, p. 113). During the analysis process, of course, the researcher's own position should also be reflected. Since the author of this study is a *white* German, has no own experience with migration and is not translocally embedded in different cross-border contexts, she might miss certain aspects regarding the research topic during the analysis. It is therefore important to note that the qualitative analysis of the data collected, as well as the prior data collection process, are affected by the author's subjectivity and therefore the analysis represents one of several possibilities of interpretation.

At this point, it is also important to mention that analyzing biographical narratives and interpreting accounts of personal experiences and emotions require taking a processual approach to subjectivity that views people as dynamically thinking and feeling beings who experience and project shifting, sometimes even contradictory, ideas about themselves. While moving through space and time, individuals are often confronted with familiar and unfamiliar situations, experience both loving and hostile environments, and must cope with contradictions (SVAŠEK and DOMECKA, 2012, p. 107). The analysis must take into account how time and space of the narration is embedded in time and space of the interview—the event of narration (PERRINO, 2015, p. 156). Biographical narratives include the present and the future, but they mainly report on the past. It must be noted that narratives of the past are always characterized by ambiguity and incompleteness. In order to comprehend the past, the individual must remember and recall past experiences, emotions, and images. In this process, memories are reproduced selectively, often incompletely, and perhaps reconstructed erroneously (EVANS, 2013, p. 21), also because it is not possible to tell everything and therefore the interviewees have to report their experiences in a condensed form (SVAŠEK and DOMECKA, 2012, p. 108).

Insights from Migrant Biographies on the Concept of *Heimat*

4

In the following, the presentation of the results is oriented along the biography of the participants. It is noteworthy that the factor of time plays an important role. On the one hand, the age of the participants should be mentioned here. The age is, for example, related to the wealth of experience of the participants. Apart from this, the experiences were made at different points in their lives. While three participants (IP1, IP2 and IP5) migrated with their families in early childhood, the others migrated at different times in their adulthood. And on the other hand, at the time of the interviews, the participants were at different points in their lives and had lived in a place for different lengths of time. The period of residence, as well as the time actively invested in creating *Heimat*, plays a crucial role in the formation and renegotiation of an individual's sense of *Heimat*.

Taking the factor of time into account, section 4.1 will focus on the initial factors that orient and, in some cases, decisively shape the individuals' sense of *Heimat*. These initial factors refer to experiences in childhood and adolescence, the places of growing up and socialization. In section 4.2, the influence of initial migration experiences on familiarization and embeddedness in the new environment will be elaborated. In this context, initial migration experiences refer to the experiences between the moment of physical arrival in the foreign environment and the feeling of having arrived on a long-term basis. The following section 4.3 is not to be considered separately from the previous chapter, as it will deal in particular with the process of *Heimat*-making and the experiences linked to it. On the one hand, the strategies that the participants pursue or have pursued in order to orient themselves in the new environment and to embed themselves will be examined (section 4.3.1). And on the other hand, the trans- and plurilocal practices for maintaining embeddedness and connectedness to a distant local context will be addressed (section 4.3.2). Subsequently, section 4.4 will present how the participants renegotiate their sense of *Heimat* in the course of

J. Andel, *Sense(s) of Heimat*, BestMasters, https://doi.org/10.1007/978-3-658-38985-7_4

their migration experiences. In particular, the different manifestations of 'being in-between' (section 4.4.1) and the understanding of *Heimat* as an ideal condition for self-expression (section 4.4.2) will be highlighted.

4.1   Initial Factors that Orient Individuals in their Sense of *Heimat*

For all participants, certain characteristics of places of growing up—the 'original spaces' (PROSHANSKY et al., 1983, p. 64)—can be identified, that both shape and orient the individual sense of *Heimat*, and thus serve as references. These references can be described as initial factors. They are interrelated but can be roughly divided into physical characteristics of the places, familiar practices, social relations, and formative experiences. These initial factors become especially visible through comparisons between 'here' and 'there'. Many participants only became aware of the meaning of these factors through the (temporal) distance. The meaning and emotional attachment express, for example, in feelings of missing and homesickness [*Heimweh*]. Often, therefore, attempts are made to compensate for certain factors and to integrate certain elements into everyday life.

Physical characteristics of places
Very prominently, many participants refer to physical characteristics of places, such as the familiar landscape (IP1, IP2, IP4, IP5, IP7, IP9, IP10, IP11, and IP12), climatic conditions (IP2, IP3, IP4, IP5, IP9, IP11, and IP12), or other sensory perceptions such as smells and sounds (IP4). These physical characteristics represent a positive reference as the participants have become accustomed to them over a long period of time and they are thus perceived as familiar. This positive reference is reflected in the fact that familiar physical characteristics are compared with the characteristics of unfamiliar places. On the one hand, similarities can evoke (positive) associations, as is visible in the statement of IP10, for example:

"Wenn ich sowas irgendwo sehe, zum Beispiel auch in Deutschland, dann denke ich immer an einen Ort, woher ich komme. Weil bei uns ist es so, [...] also ich bin geboren in einem Ort in der Nähe von der Ostsee. Und da ist so ein Kiefernwald entlang der Ostsee. Und jedes Mal, wenn ich sehe, ich denke, ok hinter diesem Kiefernwaldstück muss ein See, also ein Meer sein (lacht)" (IP10, #00:00:23-#00:00:59), "[...] also jedes Mal, wenn ich sowas sehe, denke ich immer an Zuhause, irgendwie, an meine Heimat" (IP10, #00:01:14-#00:01:29).[1]

And on the other hand, these comparisons also often refer to differences, emphasizing the unfamiliar and identifying it as not being a part of *Heimat*, as is illustrated, for example, in the statements of IP3 and IP12:

> "I don't feel like it's my Heimat when it comes to weather [...]" (IP3, #00:05:52-#00:07:06). "In the winters here I feel really really handicapped in Germany. It limits what I can do and what I'm comfortable doing" (IP3, #00:29:35-#00:30:49).

> "Habe aber halt auch hier festgestellt tatsächlich, Heimat hat für mich sehr viel mit der deutschen Landschaft auch zu tun. [...] So Burgen und unsere geordneten, also unsere Felder, also, weil ich war immer super viel draußen im Feld und Wald und so" (IP12, #00:02:48-#00:05:56). "Und ich kenne einfach viel Büsche, Pflanzen, die Tierwelt und das ist hier alles anders. Wir sind hier in der subtropischen Zone im Norden von Neuseeland. [...] Es gibt hier so viele verschiedene Tiere und Pflanzen, die es nicht in Deutschland gibt. Und ich fühle mich nicht heimisch. Es sind nicht meine Wälder, sozusagen" (IP12, #00:06:17-#00:09:35).

In such comparisons between the familiar and unfamiliar characteristics of places the participants not only identify differences or similarities, but also express preferences, as is apparent, for example, in the responses of IP9 and IP4:

> "And also the weather here is terrible in Germany (laughing), whereas back in my country, in my city, like we get over 250 sunny days" (IP9, #00:08:33-#00:09:39).

> "Ich vermisse mein[e] kanadische Heimat immer mehr im Sommer—ich bin in ein "beach town" aufgewachsen und ich muss immer daran denken, wie schön es war bei heißem Wetter meine Freunde zu treffen am Strand und den ganzen Tag am Wasser abzuhängen. Ich wohne jetzt am Rhein, aber es kann einfach nicht verglichen werden. Am Wochenende waren mein Mann und ich Pommes essen am Rhein, von einem Imbiss, man sitzt dann [auf der] Picknickbank direkt neben das Wasser. So schön wie es ist—die Pommes sind wirklich lecker—muss ich IMMER kommentieren, dass die "beach fries" vom Imbiss am Strand zuhause doch einfach viel besser sind, und sowieso auch wie es ist, auf dem Handtuch zu sitzen mit den Füßen im Sand [...]" (IP4, 15. July 2021).

---

[1] The quotes from the interviews and the online diaries are presented in their original wording. Minor corrections and adjustments are indicated by square brackets. No translation is provided, as this could misrepresent the statements of the participants. The time marker of the quotes always refers to an entire section of speech in the interview in which the statement can be found. Quotations from the diaries are marked with the date of entry.

From these comparisons, it is clear that the physical characteristics of the places of growing up have a strong influence on the individual's sense of *Heimat* and orient the individual, whether as a reference point for comparison or in terms of preferences. The familiar landscape image is predominantly perceived as positive and is often associated with positive memories. In order to appropriate an unfamiliar environment and thus make it a familiar place, it takes time, on the one hand. For example, IP4 notes in her diary entry that she always notices the different birdsongs (as well as other ambient sounds) in unfamiliar environments. She interprets her feeling of no longer perceiving the sounds of birds as foreign and exotic, but comforting and familiar, as "[…] a sign of you feeling really comfortable where you are" (IP4, 25. July 2021). On the other hand, active measures can also be taken to partially compensate for the absence of certain factors. IP3, for example, uses an extra heater in his apartment in order "[…] to make the ambient weather better" (IP3, #00:32:07-#00:34:24). Other participants feel the urge to travel regularly to the country where they spent a large part of their childhood. In addition to visiting friends and relatives, IP1, for example, always wants to visualize [*vor Augen führen*] the landscape (mountains and sea) in which she grew up.

Familiar practices
Familiar practices represent another initial factor that orients the participants in their understanding of *Heimat*, serves as a reference for comparisons, and is partly integrated into their everyday life. Here, reference is made particularly frequently to language (IP1, IP2, IP4, IP5, IP10, IP11, and IP12), food (IP3, IP4, IP5, IP6, IP7, IP8, IP9, IP10, IP11, and IP12) and other culture-related practices, such as lifestyles and social interactions (IP3, IP6, IP7, IP8, and IP12). Language (mother tongue) is explicitly described by half of the participants as an important component of *Heimat*. For IP11, for example, language represents the most important aspect of *Heimat*. As the importance of the mother tongue as well as literature was always emphasized in his family and thus was an important part of his education and socialization, for him "everything in the background is connected with language" (IP11, trans., #00:01:37-#00:02:48). On the one hand, it can be observed that language is associated with identity. And on the other hand, for many, the mother tongue also stands for familiarity, security and comfort. This is expressed, for example, in the fact that some participants say that they can communicate more easily in the language with which they grew up and were educated, because they can express their feelings better and there are fewer misunderstandings. For example, IP10 says that she can express herself in German, but relaxed speech is only possible in Lithuanian—her native language. When she

meets people with whom she shares the same mother tongue, she says, it evokes a relaxed feeling in her. The importance of language and speaking the familiar language for people's emotional well-being can also be seen, for example, in the diary entry of IP12. She is frustrated that she can neither speak her mother tongue nor the language of the country where she lives properly in her everyday life. The feeling of being 'lost in translation' is a burden on her. Therefore, she feels more comfortable and freer in 'multicultural environments' due to similar experiences.

Furthermore, frequent reference is made to food and related practices. Like language, food is an important part of everyday life and is associated with familiarity and security. Familiar dishes of food, its taste, and food practices often serve as references for comparisons. Similarities are seen as positive, while major differences tend to be seen as negative and preference for the familiar is expressed, as is evident, for example, in the interviews with IP4, IP8 and IP9:

> "[…] Frühstück ist eine Mahlzeit, eine einfache Mahlzeit, aber es unterscheidet sich sehr viel. Ich habe mich nie daran gewöhnt, Brot mit irgendwie Fleisch und Käse zum Frühstück zu essen. […] Abendbrot ist mir noch viel fremder! (lacht) […] Ich komme nicht damit klar. Das ist nicht ok (lachend). Ich will eine warme Mahlzeit. So kenne ich das. Das brauche ich auch" (IP4, #00:40:45-#00:44:32).

> "I miss a lot the food variety. I lost a bit the fun of eating (laughing). I really love Portuguese food or Portuguese from where I come from" (IP8, #00:22:03-#00:23:16).

> "[…] the food is way more different now. Especially the Bäckerei. Like that in our country is way more, like the taste of it is more colorful. I don't know how to say it. Here it is like, you get bread and then you get bread with salt, and then you get bread with salt with sesame. I don't know. It's so plain" (IP9, #00:18:18-#00:19:17).

Especially the absence of familiar food is seen as negative and is often a cause for homesickness. To create a familiar environment, many participants integrate familiar food into their daily routine. For example, IP3 regularly orders food for a several-day supply, "to make up for the lack of that, because [his] mom used to cook a lot of Arabic food […]" (IP3, #00.32:07-#00:34:24). Some participants express that they go to special stores occasionally, such as a German bakery in New Zealand (IP12) or a store with Latin American specialties in Germany (IP7), to compensate for the absence of the familiar. The preparation of meals, which can be considered as 'comfort food' (c.f. TROISI and GABRIEL, 2011), also reflects the formative influence of everyday practices. Several participants say that by preparing and eating familiar foods, they can create a familiar environment for themselves, thus connecting with their *Heimat* and counteracting feelings of homesickness (IP4, IP5, IP6, IP7, IP10, and IP12).

With regard to familiar practices, some participants also identify differences as well as similarities related to culture. Similarities are considered as positive because they are familiar and thus do not require much adjustment and adaptation to a new environment (IP9 and IP11). Differences, on the other hand, are highlighted more often. In such comparisons, it is often expressed that the familiar is preferred and missed. In many cases, these culture-related practices relate to social interactions. Some participants perceive certain social practices as particularly unfamiliar or irritating (IP3, IP6, IP8, and IP12). For example, IP3 describes the culture in which he grew up as more hospitable and spontaneous, whereas where he lives now "[...] it's all about making Termin, planning beforehand" (IP3, #00:15:12-#00:16:29). Two participants (IP7 and IP8) refer to going dancing, which played a larger role in their lives before the relocation, but is rarely practiced in their current daily lives. According to IP8, this is due to differences in social interactions and their interpretations:

"I miss a lot the dancing. I got used to not dance anymore so much. [...] I miss that before to be dancing without being judged or without being considered that you are flirting. That when you dance, you're just dancing. It doesn't have a second meaning (laughing)" (IP8, #00:22:03-#00:23:16).

In this statement, it also becomes clear that an awareness of the differences of certain social practices only grows through confrontation with other unfamiliar practices. Also, the meaning of habitual practices for the individual is reflected to many participants only through the confrontation with the 'other', but also through the absence of such practices and cognitive processes through spatial and temporal distancing. Although the participants adapt to local practices to some extent, some participants are not able to forgo what they are used to and prefer, and thus have to compensate by, for example, surrounding themselves with people with whom they can perform certain practices. For example, IP12, who moved from Germany to New Zealand, is bothered by the fact that in the country where she currently lives, her joy of discussion tends to be interpreted as a search for conflict and aggression. She considers social interactions in New Zealand to be rather conflict-averse. To satisfy her urge to have discussions, she occasionally meets people who have also moved from Germany to New Zealand, or talks to friends and family members in Germany via digital communications to discuss certain topics.

Social relations

Another very important initial factor is the social relations. This refers to social and emotional connections to people, such as family members and friends, who play a decisive role in shaping the sense of *Heimat*. It is especially the shared experiences that are made over a longer period of time which fill places with meaning. Therefore, memories of shared activities play a major role in the sense of *Heimat*. For example, IP5 fondly remembers sitting and eating together with her family, and thus spending time with family is central to her sense of *Heimat*. While this togetherness is not necessarily tied to a specific place and can take place anywhere, the experiences were made in certain places like her birthplace and former parents' residence, her aunt's residence or her grandmother's small property, which are still characterized by emotional significance for her as a result. For many participants, there is a strong emotional attachment to the places they grew up, which are associated with memories of family and friends, and therefore a connection is maintained to the people in these places and thus to the place itself (IP1, IP2, IP4, IP5, IP7, IP9, IP10, IP11, and IP12). However, this is not always the case. For some participants, due to certain experiences, such as early and frequent moving (IP6 and IP8), a strong attachment to the places of growing up is not observed. This is due to the fact that building an attachment to places not only takes time to accumulate positive experiences, but also time must be invested in building and maintaining relationships with other people. Thus, close persons, such as parents, siblings and best friends, are more important than actual places as the social relations serve as reference points. The importance of formative social relations is recognized by most participants, as with the other factors, through absence and negative feelings caused by distance. Usually, feelings of loss, missing, and homesickness relate to the social relations within the places.

Social relations as an initial factor also refer to a supposedly shared identity. This relates to the aforementioned social relations with family members and friends, but also to people with whom an identification occurs with regard to their origin, language and other shared practices. These commonalities and shared experiences create a sense of familiarity and connectedness. This leads, for example, to more or less purposeful contacts being sought and established with people in an unfamiliar environment and to networks being formed on the basis of what they have in common (IP1, IP2, IP5, IP7, IP9, IP10, IP11, and IP12). Commonalities do not necessarily have to be identified on the basis of a common origin and language, but can also be based on shared experience as a migrant, as becomes clear in the conversations with IP5 and IP12, which is why they strive to have a circle of friends as diverse as possible. In various interviews, however, it also becomes clear that certain shared experiences are only formed on the basis of socialization in the same country and thus commonalities create a bond through shared knowledge. Such shared experiences include, for example, going through the same school system, similar everyday social practices, or consuming the same food and media. For example, IP4 who identifies as Canadian states that there are certain things she can only share with other Canadians. Here, she refers to shared childhood experiences, such as eating the same candy, having the same school experiences, and watching the same cartoons, which are very different from the experiences of people who were not socialized in Canada. She also refers to the band *The Tragically Hip* which is very popular in Canada, but not outside of Canada. According to her, one can have a conversation with any Canadian about this band, as their music is a "quintessential" (IP4, #00:11:11-#00:17:27) part of Canadian identity and thus represent a national symbol that only insiders understand.

Formative experiences
Another initial factor that orients individuals in their understanding of *Heimat* can be described as formative experiences. These experiences are in-place experiences and thus are associated with places, the perception of their physical characteristics, familiar practices, and social relations within. From all the interviews, it appears that *Heimat* represents a positive reference. In most cases, *Heimat* is primarily related to positive experiences in childhood and adolescence as well as (nostalgic) memories. This is also reflected in the fact that *Heimat* is connoted by the participants with positive feelings and emotions, such as warmth, wellness, comfort, familiarity, security, safety or pride. Of course, all participants also have negative memories of experiences from their childhood and adolescence. Retrospectively, however, these negative experiences play only a subordinate role for many participants, since they prefer to remember the positive things of the past. Even though negative experiences are also considered as a part of childhood and adolescence, they are largely separated from the personal concept of *Heimat*. The exclusion of negative experiences thus also helps individuals to define what is not *Heimat*.

The exclusion of negative experiences from the understanding of *Heimat* becomes particularly clear in the interview with IP3. In contrast to the other participants, he reports predominantly of negative experiences, which are mainly based on social, religious as well as familial expectations, which he could not and did not want to fulfill. This peer pressure in the country where he grew up had a strong negative impact on his mental health. Although there were also positive moments with the family, he associates his growing up with "a lot of pain" that he "was just surviving" (IP3, #00:07:42-#00:08:43). It was only through a trip to another country to find out what life could be like without these barriers, as he describes them, when he realized that "it was all an environmentally induced psychological problem" (IP3, #00:03.06-#00:05:41), as he experienced a release of all pressure there. Due to the negative experiences over a long period of time and the confrontation with alternatives, he decided to leave the country. Also because in his eyes "the other option was suicide" (IP3, #00:08:50-#00:09:3). Because of these strongly negative experiences, which contrast with all the positive feelings and emotions typically associated with *Heimat*, a forward-looking understanding of *Heimat* emerges. The past including the places of childhood and adolescence are not part of his sense of *Heimat*. Consequently, an emotional detachment from the places can be noted. Nevertheless, it can be stated that these negative experiences shape and orient the understanding of *Heimat*, since the personal feeling of *Heimat* is also always related to what is not *Heimat*.

How strong the social and emotional attachment to places is depends not only on the quality of the experiences, in the sense of whether they are positive or negative, but also on the factor of time. In order to develop a sense of *Heimat*, it takes a very individual but rather long period of time spent in places. As mentioned above with reference to the interviews with IP6 and IP8, frequent moving and thus several short stays can prevent the development of a strong attachment to places. IP6 tells in the interview that she moved out at an early age—she was sixteen—and since then has moved frequently and also lived in other countries. Due to the frequent moves, temporary stays and distances, only a few close social contacts could be maintained. Because of these experiences, it was difficult for her to locate *Heimat*. In addition to social contacts, she also associates symbols such as a cookbook with dishes from the region of Styria in Austria where she grew up and where her parents live, as well as her passport with *Heimat*. The passport as a symbol of *Heimat* is particularly revealing about her understanding of *Heimat*. On the one hand, this passport is her constant 'companion' and records her stays in North and South America. And on the other hand, this official document testifies to her national belonging and identity. Thus, there are certain social and emotional attachments that represent something familiar, function as reference points and orient her in her sense of *Heimat*. A strong emotional attachment to specific places that are considered as a part of *Heimat* cannot be identified. This is similarly evident in the interview with IP8. She reports that due to the problematic financial situation of her single mother, she was forced to move frequently as a child. During all this time, however, she attended the same kindergarten, school and later university in the city center while she lived on the outskirts of the city. Relationships with friends could be built up and maintained as a result. However, no strong bond was built up with the places of residence. She reports that the great spatial and temporal distance that she had to overcome every day on the way from the outskirts of the city to the kindergarten, school and university in the center led her to vividly remember public transport. These experiences of inconstancy thus also shape the sense of *Heimat*, as becomes clear in the interviews with IP8 and IP6. The sense of *Heimat* is oriented mainly and more overtly than in the case of other participants to emotional bonds with important people. Only minor personal significance is ascribed to places from childhood and adolescence.

4.2   Influence of Initial Migration Experiences on Familiarization and Embeddedness

As already mentioned, the participants migrated at different points in their lives. The reasons and motivations for migrating to another country also differ, although similarities can also be observed. Since different motivations and (economic) framework conditions often converge, a clear categorization into voluntary and involuntary migration is not always possible, but also not purposeful. To illustrate this briefly: The two sisters IP1 and IP2 moved with their family from Germany to Greece on a long-term basis in their early childhood, which was not their decision but that of their parents. However, there was no economic predicament and the move was perceived as a "smooth transition" (IP1, #00:04:08-#00:05:13), as there was already a strong connection to Greece due to the family on their father's side. Later they moved back to Germany. IP5 also migrated early with her parents. However, her parents were forced to leave the country—which no longer exists today—due to the Yugoslavian war and moved to Germany. IP5 did not understand the situation as she was too young and only later realized the circumstances. IP3 was also in a predicament, as he was forced to leave his familiar environment in the UAE in order to escape societal expectations and eventually save his life. Although this was a forced migration, his flight experiences differed significantly from those of many other refugees due to his secure financial status. He also clearly perceives this as a privilege. The fact that economic factors always play a major role in migration is also reflected in the interview with IP9. He was not directly forced to leave North Macedonia where he grew up, but he was not satisfied with the economic circumstances in his country and was looking for better living conditions and economic opportunities to be able to provide for his family. IP10 and IP11 were also looking for better educational opportunities, initially leaving Lithuania only for university education, but then realized that their country of origin offered fewer career opportunities and other factors were also more appealing in Germany where they live now. IP7 and IP8 also initially left Colombia (IP7) and Portugal (IP8) for a period of study in Germany (IP7) and Austria (IP8). They chose the destination mainly because of financial reasons such as lower tuition fees, and later found better career opportunities, so they decided not to return. IP4, IP6, and IP12 also initially chose a temporary stay for study (IP4 and IP6) or work-and-travel (IP12) in Germany (IP4), the USA (IP6), and New Zealand (IP12). However, the short-term stay turned into a long-term stay as all three met their current relationship partners.

The motivations and circumstances of migration obviously play a role to the extent to which individuals can appropriate and thus familiarize themselves with a new environment. The degree of embeddedness in a local context is also influenced by motivations and circumstances. However, initial migration experiences have a much greater influence on familiarity and embeddedness. It became clear during the interviews that it is especially the initial experiences that significantly influence the development of the personal sense of *Heimat*. This can be in a positive direction, so that feelings of having arrived, belonging and being accepted occur, which facilitate the development of a sense of *Heimat*. These initial experiences, however, can also negatively influence the development of a sense of *Heimat*. Above all, experiences of exclusion and uncertainty prevent the development of positive feelings associated with *Heimat*. Three main aspects emerge from the interviews that influence the participants' familiarization with and embedding in the new environment. The first is the prior knowledge, which affects the experiences and thus the formation of a sense of *Heimat*. Second, feelings and situations of uncertainty due to the legal status and other barriers to long-term residence play a role, which can be summarized as future prospects. And third, the diverse social encounters have both positive and negative effects on the development of a sense of *Heimat*.

Prior knowledge

For many participants mainly their language skills were decisive for their first experiences in the new/different environment. Few of the participants moved to a country in whose official language they were able to communicate confidently before the move. Depending on the motivation for the move, prior linguistic knowledge also varies—either not existing at all, acquired through language courses beforehand or locally, or already acquired over a longer period of time through family bilingual upbringing, language acquisition during school, or short-term stays abroad. Through the reported experiences of the participants, it can be stated quite banally that the lower the language skills, the more difficult it is to orientate in and get accustomed to a new environment and more negative experiences may occur. For example, IP1 reports that because of her bilingual upbringing in Germany, she had no problems meeting other children, making friends, and going to school in her new place of residence in Greece. She could not only speak Greek, but unlike other children in elementary school, she could even write, which is why she was at the top of her class in Greek lessons. This prior knowledge made her socialization at her new place of residence much easier, which is why she also felt very comfortable, as she tells in the interview:

> "Also das war für mich wirklich so ein reibungsloser Übergangsprozess" (IP1, #00:04:08-#00:05:13). "Und ja, ich kann mich einfach nicht daran erinnern, dass für mich irgendetwas schwer war oder dass ich mich da nicht wohlgefühlt habe" (IP1, #00:07:10-#00:09:56).

On the contrary, for example, IP5 had no knowledge of the German language when she fled to Germany with her parents as a small child from Yugoslavia. Not only she but also her parents had great difficulties in communicating, which is why it was difficult to orientate herself in the new environment and to establish relationships with local people. Since there were no integration and language courses for refugees in the 1990s in Germany—these were first introduced in 2005 (BMI, 2020)—they were left to learn the language in their everyday life by memorizing and repeating individual words and phrases. IP9 had a similar experience, initially coming to Germany from North Macedonia on a short-term basis to work, also to find out if Germany would be the right choice to emigrate to. He also came with no prior knowledge of the language and tried to learn the language in his everyday work environment in a fast food chain, as he describes:

> "Well, basically when I came here the only words I knew were Ketchup, Mayo and like, yeah, Gurken. [...] I would say like there was an app called Duolingo that helped me a lot. Like, I would go home and I would just memorize all the words that the people told me but I did not understand and tried to translate them as quickly as possible and just try to memorize them" (IP9, #00:14:06-#00:15:18).

Now that he has moved to Germany on a long-term basis and changed jobs, he does take language courses, but still does not feel confident communicating in the language and prefers English when given the choice. IP3 and IP8 also draw on a third language like English in their everyday lives, especially when meeting new people or in the work environment. Both of them also had no prior knowledge of the German language when they migrated to Germany (IP3) and Austria (IP8), but can now communicate in German and understand most of it, but nevertheless they prefer English. That many people feel uncomfortable communicating in a language they are not (yet) confident in, is partly because, as described once above in regard to familiar practices, the limited vocabulary means that only limited communication is possible. Not being able to express oneself freely causes a feeling of discomfort. This feeling can also be evoked by other people's reactions. All participants who do not speak the official language in the country they live in without an accent or without grammatical errors report various situations in which they were pointed out to their deficit or were asked about their country of origin. These reactions, depending on the person and situation, are either perceived as positive expressions of support and interest (IP4, IP7, IP10, and IP11) or as degrading and exclusionary experiences (IP3, IP4, IP5, IP7, and IP12).

Prior knowledge that influences the initial experience and the process of familiarization and embedding also includes a person's knowledge of an environment before the long-term move and the connections that already exist, for example, through local social connections. Half of the participants initially left the country they grew up in for a short period of time to study (IP4, IP7, IP8, IP10, and IP11) or to work (IP9), and then decided to stay for the long term for different but also very similar reasons. None of them had visited the country before and there were very few to no local social connections, unlike other participants (IP1, IP2, IP3, IP6, and IP12). While some participants were able to adjust and orient themselves quickly, others had to completely reorient themselves and were dependent on support. The importance of positive experiences of a welcoming atmosphere and the support of local people is shown, for example, in the conversation with IP8:

"[…] I decided to come to Graz. I didn't know anything about the city. And it was surprisingly easy (laughing). Even with the difficult financial situation I had I could manage to do everything. The university here was very helpful. [...] Maybe that's why I have so much good memories or such a good feeling about Graz. And maybe that's why I have stayed for these 11 years. Maybe because it started so easy. And it [has] been always very welcome in every year that I am spending here, from Erasmus to deciding to finish my studies here. I was also very well accepted. And I got help from the teachers and from university to learn the language. And then also after finishing university I was not having troubles to find work, because I got a lot of also help to get connections from my teachers. And I am making career here, still (laughing). So, it seems like it was the best decision to stay. It seems like it was made for me to stay here, because everything runs ever so smooth" (IP8, #00:14:05-#00:16:36).

In this statement from IP8, it is evident that the support of other people can facilitate and ease the transition from a short-term stay, such as during a semester abroad, to a long-term stay with career opportunities. The opposite experiences of IP4 and IP9 also seem to confirm this. Due to their initially short-term stay in the context of a semester abroad (IP4) and a temporary job (IP9), both initially had a kind of safety net through a community of fellow foreign students and work colleagues as well as the framework conditions of their stay. However, the transition to a long-term stay was difficult for both of them. Once the safety net vanished and they were confronted with everyday situations, a "culture shock" (IP4, #00:24:35-#00:40:36) set in, as IP4 describes it. Both describe this as a difficult transition phase:

"Well, I was basically on and off. I was not sure, if I wanna like stay. The first year especially, because I was like literally alone here. There was no like someone to help me with Ausweis or health insurance or like translating my driving license. Nothing. And I was like clueless. And it was depressing" (IP9, #00:15:48-#00:17:08).

"Das ist eben, wie ich meinte mit dieser schwierigen Phase in Freiburg nach dem Austauschjahr und zum ersten Mal wirklich alleine sein. Ich hatte erst keine community, keinen Freundeskreis, keine anderen Ausländer, die mich unterstützen. Ich hatte nur Marks [her husband] community. Und ich musste mich da irgendwie integrieren. Und wenn ich mich da nicht zurechtgefunden hätte, oder wenn ich mich mit denen nicht wohl gefühlt hätte, hätte ich nie diese Gefühle von Heimat haben können. Dann wäre ich nur froh davon weg zu kommen" (IP4, #01:07:23-#01:10:31).

From the interviews with other participants, who had prior knowledge, for example, by being familiar with the country or certain places to some degree before the move due to previous visits and family-related connections (IP1, IP2, IP3, IP6, and IP12), it is evident that this can foster and ease the process of embedding and the formation of a sense of *Heimat*. For example, being already familiar with the surroundings and having social relationships can make it easier to navigate through daily life and to embed oneself in a local context by the facilitation of access to bureaucratic and institutional matters, job opportunities or social networks. However, since there are also other factors and initial experiences that influence these processes of familiarization and the degree of embeddedness prior knowledge alone is often not enough.

Future prospects
Another important aspect that shapes the experiences of migrants and contributes significantly to the familiarization and embeddedness are the future prospects. When people intend to stay in a country only for a limited time, for example, as part of a study program, for professional reasons or for other clearly defined temporal activities, their experiences in daily life in a country tend to be different

from that of people who live there on a long-term basis. This is evident, for example, in the statements of IP4 and IP9 quoted above, who experienced a difficult transition phase from a short-term to a long-term stay, as a new safety net (employment, accommodation, insurances, community, social networks, etc.) has to be established and new everyday challenges (language barriers, bureaucracy, establishing social relationships etc.) have to be overcome. From many interviews it is clear that positive future prospects, in the sense of a promising plan of long-term stay, not only create a sense of stability and thus a sense of *Heimat* can develop, but also depend on a sense of stability and the overcoming of uncertainty. Decisive factors for the intention to move from one country to another in the long term and for positive future prospects are, on the one hand, the legal status (IP3 and IP12), the possibilities in terms of professional career and education as well as the systems such as health care and social security (IP1, IP2, IP3, IP6, IP7, IP8, IP9, IP10, IP11, and IP12), but also a romantic relationship and (plans to) starting a family (IP4, IP5, IP6, IP10, IP11, and IP12).

Some participants aspire for a citizenship in the long term (IP3 and IP7), as it enables participation, for example in the form of the right to vote, but it is also connected with certain rights and a feeling of security. However, citizenship is not an option for everyone, as dual citizenship is not an available option for those participants who do not want to give up the citizenship they already have in their country of origin (IP4 and IP11). For some participants, the legal status in the form of citizenship (IP3) or a permanent residency visa (IP12) plays a very important role in their sense of stability, security and belonging. This is illustrated, for example, in the following statement by IP12, who has been living in New Zealand for over three years with a break of several months in Germany:

> "[...] es spielen hier so viele Dinge eine Rolle dabei heimisch zu sein. Also jetzt habe ich mein Residenz-Visum sozusagen und das ist auf jeden Fall ein großer Schritt. Und das wird sich wahrscheinlich so in den nächsten Monaten zeigen, wie viel Wirkung das tatsächlich auf mich hat. Und seitdem habe ich schon das Gefühl, dass ich mich ein bisschen mehr zugehörig fühle" (IP12, #00:06:17-#00:09:35).

For most participants, however, career opportunities and other safeguards play a much larger role in their decision to stay and their path to settling down for the long term. For example, the sisters IP1 and IP2, who were both born in Germany, moved to Greece with their family as children and are now embedded in both countries, however, see their future mainly in Germany because of more promising career options. For IP7 and IP8, who both found quickly a satisfactory job after their studies, the decision to shift the center of their lives to another country and to stay in the long term was also easy. This is also because the limited opportunities and economically poorer prospects in the country where they grew up can be seen as a push factor. This is also evident in the interview with IP9, who saw no future in his birth country North Macedonia and moved to Germany in search of better opportunities to earn money. Unlike the participants who first came to study in another country and then found a well-paying job, he had to start out in a precarious employment situation. Experiences, which are shaped by a person's economic and social position, can have an inhibiting effect on the development of a sense of *Heimat*. IP9 describes his experiences and current situation in the interview as follows:

> "[…] I'm in the lower ranks of society, I would say. It might sound dumb, but I'll explain (laughing). Like, us [North] Macedonians we would start off in McDonalds, for example. That's how we all get to work here in Germany. And you're in the same circle of people. Basically all [North] Macedonians or like from the Balkans, Serbia, Bosnia and so on. So, you don't get to feel the life and the surroundings of Germany, for example. So, I haven't felt it yet. Hopefully, like soon, soonish, when I start— I plan to start an Ausbildung—I would connect with more people and therefore I'll maybe start feeling it, like the Heimat-feeling (laughing), I guess" (IP9, #00:06:00-#00:07:54).

His statement reveals that a job not only provides (financial) security, but also embeds an individual in the society and a local context by enabling participation and establishing social relationships. It is especially the opportunity to connect with other people, build relationships, be part of a community, and establish networks that emerges in many interviews as essential for positive future prospects and plans as well as the development of a sense of *Heimat*.

In addition to establishing friendships, engaging in a romantic partnership in particular plays a crucial role. For example, IP4 and IP12 stated that their current relationship partner (IP12) and husband (IP4) was the decisive reason to stay. Through a romantic relationship, individuals not only become emotionally attached to a person and a place, but also gain access to local communities, networks and experience. On the one hand, they receive practical support to orient themselves, establish a daily routine and overcome various barriers, and on the other hand, they are offered emotional support to cope with the transition phase to a long-term residence and to process negative migration experiences. The partners of IP4 and IP12 thus can be seen as fixed anchor points that contribute significantly to their embeddedness in the local context. In addition, a romantic relationship influences future plans by raising new questions regarding where to reside or whether and how to start a family (IP4, IP5, IP6, IP10, and IP11).

These considerations are perceived as very profound by those participants who are in a committed relationship, are engaged or married, or already have a child together. All of them believe that a family and especially a child needs a stable environment in which to grow up. Here, of course, they also refer to their own experiences.

Social encounters
Social relationships play an essential role in the formation of a sense of *Heimat*. This is evident at different points in all interviews. Before and whilst individuals establish new social relationships in an unfamiliar environment, various social encounters occur, which have an important initial effect on the familiarization and embeddedness of a person in a local context. On the one hand, these can be positive social encounters and interactions which are characterized, for example, by goodwill, support or love, and thus create a welcoming atmosphere. Such initial positive experiences facilitate the arrival and the process of settling, encourage engagement and the process of embedding, and help to create a sense of acceptance and belonging, as already described above in different examples. Negative social encounters, on the other hand, are also particularly formative as they negatively influence the emotional well-being and thus interfere the development of a sense of *Heimat*. From the interviews, it emerges that social encounters are mostly perceived as negative when otherness is emphasized, thus belonging is questioned and the individual is excluded. Otherness, and thus the localization of an individual in a foreign and different local context, is mainly defined by people who are embedded in the local context. This shows that the sense of *Heimat* is not only the result of an individual process, but is also strongly dependent on other (groups of) people. Thus, recognition and belonging is an important part of *Heimat*. The interviews reveal that the participants' otherness and foreignness is mainly defined and evaluated by their language skills, for example by using a third language for communication such as English, a noticeable accent or grammatical errors, as well as by other characteristics such as their name or appearance. Such social encounters, where otherness is emphasized, for example by asking about the origin, are not perceived by all participants as an expression of exclusion, but rather as a legitimate interest by the questioner. Especially for participants whose identity is strongly linked to their country of origin and who identify themselves, for example, as Canadian (IP4), Colombian (IP7) or Lithuanian (IP10 and IP11) living in Germany, highlighting characteristics marked as foreign, such as language or name, does not represent an exclusion. From the perspective of these participants, their 'otherness' does not contradict their belonging or sense of *Heimat*. However, this perception also depends on the situation in

which such an encounter takes place, also in what way the otherness is communicated and at what time in life this is happening. IP4, for example, reports about her difficult phase of transition from a short-term to a long-term stay in which she felt like an outsider for a long time due to her language skills. One situation in particular, which she says she cannot forget, affected her severely on an emotional level. She refers to a conversation she had at that time with her still limited language skills, which led to problems in communication. However, it was not until another person intervened and attempted to clarify the situation, but treated her 'like a child', that she felt she was not being taken seriously and her efforts were disregarded:

> "Und der hat uns dann angeguckt und ihr dann erklärt: 'Ja, die [IP4], sie kommt ja aus Kanada. Sie kann nicht so wirklich Deutsch' und so wirklich große Augen dabei gemacht, langsam gesprochen und übergestikuliert. Und das war für mich wirklich... Ich war so traurig. Ich hätte weinen können. Weil ich mir so viel Mühe gegeben habe mich auf Deutsch zu unterhalten. Und zu diesem Zeitpunkt auch wirklich viele Verbesserungen gemacht habe. Und dann redet der, als ob ich 5 wäre. Und 'Ja, die is eh ein bisschen blöd und kann das nicht'. Boah, das war heftig. Solche Sachen am Anfang, das ist schwierig" (IP4, #00:24:35-#00:40:36).

Also IP12, who does not consider the country she has been living in for three years as her *Heimat* yet, but is involved in many ways in order to embed herself and create a *Heimat*, experiences social encounters that emphasize her otherness as negatively affecting her sense of belonging. Her accent and country of origin often function as a distinguishing feature for insiders, marking her as an outsider and legitimizing further questions, such as regarding the length of her residence so far. In addition to language skills, other characteristics such as the time spent in a place so far or the number of acquaintances can be decisive for the acknowledgement of efforts and the affirmation of belonging, as can be seen in the following statement by IP12:

> "Und wo ich mich eigentlich komischerweise immer ein bisschen exkludiert fühle ist, wenn mich Leute nach meinem Akzent fragen oder wenn die ansprechen, dass ich Deutsche bin, weil es dann irgendwie sofort irgendwie, weil dann kommt die Frage: 'Ja, wie lange bist du denn schon hier?' Und dann sage ich: 'Naja, drei Jahre'. Und ich bin halt noch neu, weil drei Jahre ist halt noch neu in einer sehr ländlichen community wie hier. [...] Und sobald mich Leute danach gefragt haben, fühle ich mich irgendwie ausgeschlossen. Und dann habe ich das Gefühl, ich muss doppelt arbeiten und erwähnen, dass ich diese fünfzig Leute hier kenne, um dann irgendwie wieder dazuzugehören oder auch eine Meinung zu haben, oder Meinung haben zu dürfen" (IP12, #00:55:55-#00:58:59).

The identity of an individual can also be linked to the sense of belonging and the sense of *Heimat*. While for most participants the identity is tied to the country of origin and is therefore clear to them, the self-conception is more complicated for people who, for example, have been embedded in different countries and language areas due to their upbringing in their family. Social encounters and the recognition of other people play an important role here. Highlighting certain identity characteristics that are marked as different can have an exclusionary effect on individuals by foregrounding otherness and defining the individual by it. IP1, whose *Heimat* is in both Germany and Greece and whose identity is German and Greek through two native languages, family connections in both countries and her life in both countries, nevertheless feels a stronger sense of belonging to her Greek side. She explains that this is because in Greece she is accepted as a Greek through her name, language skills, and biographical incorporation. In Germany, on the other hand, she is repeatedly reduced to her 'Greekness' and thus assigned to a foreign local context and marked as other.

Some interviews, however, also reveal social encounters that exhibit an explicitly exclusive and discriminatory character. Such rejecting and hostile encounters are based on prejudices as well as xenophobic and racist attitudes. The effects of such negative social encounters on the individual range from situational feelings such as irritation and discomfort to lasting psychological distress. It also depends on when and how often such encounters occur in an individual's life. Such situations can have a negative impact on the sense of belonging and thus on the development of a sense of *Heimat*. But such negative encounters can also be perceived as single irritating situations, therefore they are not emotionally stressful in the long term. This becomes clear in the interviews with IP3 and IP6. Such situations can, for example, as IP3 reports, relate to communication with people at public authorities, where he was rejected on the phone and rudely told not to call again until he could speak German properly. This discriminatory experience surprised IP3, as he expected more professionalism from a government agency. In general, however, he describes his experience and encounters as positive. Social encounters, which are characterized by prejudice and hostility, can also take place within communities, circles of friends and the family. IP6, for example, reports that she initially encountered great mistrust in her fiancé's family. The family of her boyfriend assumed that IP6 was only in a relationship with him and wanted to marry him in order to get a green card and to be able to immigrate to the United States. In contrast to such individual encounters, which are merely perceived as irritating, in particular hostile and violent encounters, which can also extend over a longer period of time, can have a lasting negative impact on the sense of belonging and shape and even prevent the development of a sense of

Heimat. As an example of this, particularly the interview with IP5 stands out, in which she reports on her hostile encounters, which she already experienced as a child. She remembers that she was part of one of only two foreign-born families living in a small town in Germany in the 1990s. Because of language barriers, she could hardly socialize with other children, and as a foreigner, she felt like an outsider. However, she experienced the conflicts with other children in the neighborhood as particularly emotionally stressful. These conflicts were especially initiated by one child who bullied her with insults due to her hostile attitude towards people originating from Yugoslavia. But it did not only remain at psychological violence, because the child also incited other children against her, and they also attacked her physically, for example by throwing pebble stones at her. These negative experiences affected not only the emotional state of IP5 but also of her parents, who were powerless due to their limited possibilities to support their child. Attempts by the mother to contact the bullying child's parents were not very successful either, as she had difficulties communicating in German and the child's parents did not want to admit their xenophobic attitude. Because of these exclusionary and emotionally stressful experiences, IP5 was never able to build up a sense of *Heimat*. The positive memories are limited exclusively to the parental home. The place where she grew up and went to school is not part of her *Heimat* today, as she does not feel any (positive) emotional connection to this place. Other negative experiences in other places in Germany, such as the search for an apartment, which turned out to be difficult because of her last name, also shaped her sense of *Heimat* in such a way that it is difficult for her to perceive Germany as her *Heimat*. She explains this difficulty in defining *Heimat* and linking it to places in Germany with the fact that she was often not accepted and thus never felt she had arrived. This example shows most impressively how important positive experiences in the sense of being accepted, supported and considered as part of a community are for the formation of a sense of *Heimat*.

4.3   Processes and Experiences of *Heimat*-Making

The interview partners were at different phases in their lives and in their migration experience at the time of the interviews. Some of the participants have been living (mainly) in a country for several years and have developed a feeling of having arrived and settled. They, thus, have already gained many personal experiences of their process of creating their *Heimat* (IP1, IP2, IP4, IP7, IP8, and IP11). Other participants are currently still in the process of creating and defining their *Heimat* (IP3, IP5, IP6, IP9, IP10, and IP12). How much time is required for one to create and negotiate *Heimat* is very individual. In general, however, it can be observed that the process of *Heimat*-making requires several years. In the following, the different strategies that the participants have pursued or are pursuing in order to embed themselves in a new environment, to appropriate it by making it familiar, and thus to create their *Heimat* will be examined (section 4.3.1). Subsequently, the trans- and plurilocal practices, that maintain the connection to the social relations and places in the country that was left and thus embed the individual across borders in more than one local context, will be described (section 4.3.2).

4.3.1   Strategies to Embed into a New Environment

All participants had to and have to more or less adapt to the new environment. As can be seen from the aforementioned, an environment consists, firstly, of the physical characteristics of places, such as the landscape or the climate. Depending on how strongly individuals are oriented towards their initial experiences in the places of growing up in terms of their physical characteristics, or to what extent these characteristics differ or resemble each other, it takes more or less time to familiarize themselves with the new environment and to develop a sense of *Heimat*. Thereby a comparison with the familiar or past references is made, even if a familiarization already occurred. In addition, individual measures are also implemented, such as certain seasonal adjustments (e.g. extra heater in the apartment) and rituals (e.g. preparing and eating certain food), in order to create a familiar and secure place or to compensate for the unfamiliar. Secondly, an environment comprises various institutions that form an essential part of everyday life, integrate individuals into a social context and enable social participation. This includes, for example, schools, universities, and workplaces. For those participants who migrated as children, initial involvement took place through education in the school system. For many other participants, initial involvement occurred through attendance at a university. However, the workplace emerges as particularly important for embedding in a new environment. Finding a job, earning money, and pursuing a career is a critical motivation for many participants to stay on a long-term basis, and is thus essential for creating a sense of security and settledness. A job functions not only as a safeguard, but also as a strategy to become part of society by contributing and socializing. And thirdly, (social) practices are also part of an environment. In a new environment, individuals have to get used to the unfamiliar practices, adjust to them or even incorporate them. Depending on whether and how much the practices differ from the familiar practices, for example in terms of language, food or social interactions, a greater or lesser adaptation is necessary. Often a public and context-specific adaptation takes place, while in the private space familiar practices are acted out to compensate for the unfamiliar, the non-preferred or the missing. The practice that is essential for embedding in a new environment is speaking the official language of the country. Overcoming the language barrier is perceived as difficult by all participants who had to learn a new language. Nevertheless, language acquisition, for example in language courses or in their daily lives, is for all participants an important part of their migration experience and process of *Heimat*-making.

Establishing social relations

All activities of embedding in a new environment take place in, depend on, or serve to establish social relationships and networks with other people. In all interviews, it becomes clear that the process of *Heimat*-making is dependent on other people who facilitate or impede it. Since a sense of belonging and emotional attachment to places is based in particular on positive social encounters, the acknowledgement by other people and shared experiences, social relationships are an essential part of *Heimat*. Thus, establishing social relationships plays an important role in orienting oneself in a new environment as well as in the process of *Heimat*-making. The interviews reveal that the participants pursued and are pursuing different strategies to establish social contacts and to build relationships. For example, some participants draw on pre-existing contacts who also immigrated from the same country, or intentionally seek contact with people who are from the same country of origin, speak the same native language, or share similar experiences in order to build relationships and form networks with them (IP5, IP7, IP9, IP10, IP11, and IP12). Especially in the early phase of the *Heimat*-making process, social relationships with people of shared identity and experiences help individuals to orient themselves. In a new and unfamiliar environment, such people also provide a familiar and secure space in which it is possible to perform familiar practices, such as speaking the native language, eating certain (traditional) food, celebrating certain festivities, and so on. Thus, for some participants, the circle of friends consists mainly of people who originate from the same country or region. IP11, for example, has built up a network of Lithuanian acquaintances and friends since his studies at a German university. He highlights that his best friends in Germany are also Lithuanians. When his spouse IP10 was asked when she felt comfortable and settled in Germany, she answered:

> "Ich glaube, das war als ich meinen Mann kennengelernt habe, weil ich dann mehr Menschen kennengelernt habe, die hier auch wohnen, die aus Litauen kommen [...]

Weil zum Beispiel in Mainz gibt [es] unter unseren Freunden ganz viele Litauer. Und das ist auch so wie eine kleine Familie, wenn wir irgendwelche Feste oder sowas haben, dann feiern wir zusammen. Und da gibt es auch eine Litauische Gemeinde, auch ein Litauisches Gymnasium sogar. Und da hat man auch so ein bisschen das Gefühl, dass es auch hier wie Zuhause ist" (IP10, #00:14:24-#00:15:06).

Similarly, to the experiences of couple IP10 and IP11, for IP9, people originating from the same country and speaking the same native language created a familiar environment and facilitated his arrival. IP9, who started working in a fast food chain in Germany with many other North Macedonians, reports that these people formed a friendly community and therefore he felt comfortable. Because of these contacts, even today, after finding another job and wanting to stay in Germany for the long term, his circle of friends consists mainly of a group of North Macedonians. Other participants, such as IP5, IP7, and IP12, seek not exclusively contact with people of shared identity and experience, but a large proportion of their friends and networks are composed of them. IP5 and IP12 report that especially the exchange with people who also experienced migration contributes significantly to processing one's own experiences and to feeling comfortable in a place. In particular, being understood, accepted and not having to explain oneself plays a major role in this. For this reason, both of them state that they feel most comfortable in places with diverse people. All participants who have established a circle of friends and networks of people who share the same national identity or similar migration experiences, as well as specifically seek contact with such people, report that certain topics can only be discussed with people who originate from the same country or have had similar migration experiences.

All participants, however, try to embed themselves in a place also by seeking contact with people who they consider as locals, because they have been embedded in the social context for a long time, speak the official language, were socialized in the country and so on. Relationships with locals are often established in everyday institutions such as school, university, or the workplace. IP9, who has had limited contact with people outside of his North Macedonian circle of friends and has been able to experience little of life in Germany, as he says, because his work environment made this difficult, plans to begin an apprenticeship in order to socialize, embed himself, and create his *Heimat*. IP12, who has lived in New Zealand for 3 years, also seeks to connect with people through her job, but also through her social commitment. To establish a *Heimat* for herself and to be recognized by others, she says, social networks are particularly important in New Zealand. Therefore, she works professionally in the field of community outreach and is also privately involved in many projects. Her goal is to build a community as well as many networks. Also, being a member of an association, especially a sports club (the participants mentioned, for example, a shooting club (IP11) or a roller derby team (IP4)), enables many participants to connect with locals through joint activities, to establish friendships, and to be part of a community.

In the process of *Heimat*-making, however, the strategy of intentionally avoiding people on the basis of similar characteristics (native language, country of origin, etc.) can also be pursued. This strategy is evident in the interviews with IP3 and IP4, although the reasons are very different. IP3, for example, who had to leave the UAE due to his negative experiences, especially in relation to other people, the emotional and psychological stress and the resulting restriction of his quality of life, has great difficulty socializing in Germany with people who grew up in a similar cultural and religious context. Due to his negative experiences and hostile encounters with Arab and Muslim people regarding his homosexuality, he is very cautious, avoids contact and only approaches such people through other people, such as friends. In order to concentrate on his new life in Germany, to become part of society and to create a *Heimat* for himself, he focuses mainly on his career and enters into only a few but selected and deep friendships. Other reasons for avoiding contact with people based on similar identity characteristics can be found in the interview with IP4, who moved to Germany from Canada more than ten years ago. She remembers keeping a distance from her Canadian student group during her semester abroad in Germany, for example, by specifically looking for a dorm on the outskirts of the city and taking many classes outside of the exchange program. She felt it was important to get to know her new surroundings in Germany, have her own experiences outside of the Canadian group, meet new people, and speak as much German as possible to learn the language. Even after her time at university, when she decided to stay, she avoided contact with North Americans. Not only because she wanted to speak as much German as possible in her everyday life, but also because she was afraid of being perceived as not conforming and confirming negative stereotypes about Americans. In order to be perceived as integrated by other people, she surrounded herself mainly with German speakers and also partly suppressed her personality, which she describes as outgoing, in order to not fulfill the image of a North American. In retrospect, she describes her behavior as overcompensation, which helped her embed herself in her social environment, learn the language, and build a *Heimat*, but also made her more aware of her own identity and caused emotional stress. Today, she calls Germany and the places in Germany where she has spent a lot of time and attached many memories *Heimat*. Her belonging no longer needs to be proven in her eyes and thus she no longer excludes having friendships with North Americans.

Contributing to society

It is also clear from several interviews that the process of *Heimat*-making is related to participation and active engagement. For some participants, it is important to contribute to society in order to be part of it. Thus, *Heimat* can be actively created and shaped. For some participants, working in a job causes the feeling of 'being a useful' and contributing part of society (IP3). Others engage in local matters and participate, for example, in local decisions by signing petitions (IP8), or volunteer in associations, initiatives and projects, using their (migration) experiences and striving to make socially relevant contributions (IP5, IP6, IP7, and IP12). As a student, IP7, for example, got involved in a political project specifically for migrants. Driven by her own experiences, she also wanted to raise awareness of political issues among other migrants. IP5, who identifies and locates herself as a European on the basis of her migration experiences, is specifically committed to superordinate European organizations with a focus on migration issues. In this way, she also strives for a *Heimat* across national borders, in which all people see themselves primarily as human beings. Engagement as a strategy to create *Heimat* becomes particularly visible in the interview with IP12. In order to establish a *Heimat* in New Zealand and to develop a sense of *Heimat*, she feels that she first has to work for her place in society. She puts a lot of time and effort into her social engagement, working on projects and building networks to 'earn' her place in society and eventually obtain a sense of *Heimat*. Before she moved to New Zealand, she was also engaged in Germany, primarily with ecological matters, but in her new place of residence, she feels a greater urge to get involved, especially in the social field. As she also deals critically with the history of the country, she tries to build a diverse and inclusive community and in particular does anti-racist work. Here she also tries to contribute with her German identity and the experiences from the history in relation to National Socialism, which also serve her as a motivation to promote anti-racism.

Some participants aspire not only to engage socially and participate at the local level, but to participate politically at a superordinate level. The right to vote is considered an important part of *Heimat* by two participants (IP4 and IP7). IP7 explains her desire to be able to vote as follows:

> "[...] ich möchte auch irgendwann dann hier wählen können, weil ich hier lebe und hier wohne. Und dann möchte ich auch, dass wo ich wohne auch ein schöner Ort ist. Das kann man nur erreichen, wenn man richtig wählt" (IP7, #00:16:27-#00:17:23).

In her statement, it is evident that participation (co-determination and co-creation) is part of *Heimat* and thus also part of the process of *Heimat*-making. In order to be able to participate politically, citizenship must be acquired. While for some people the acceptance of citizenship is a logical step towards participation and belonging, for others the legal framework creates an inner conflict. IP4, for example, would like to be able to vote in Germany as she considers political participation as important. However, accepting the German citizenship requires her to give up her Canadian citizenship (c.f. BVA, 2021). IP4, however, is not willing to give up her Canadian citizenship and thus a part of her identity. Consequently, she can neither vote in Germany, nor in Canada due to her residence in Germany. In view of this circumstance, she feels excluded and frustrated. She explains her feelings as follows:

> "Ich kann […] in Deutschland nicht wählen, was ich für eine absolute Frechheit empfinde. Ich zahle hier Steuern, ich wohne hier, ich habe mich total integriert, ich habe eine unbefristete Aufenthaltserlaubnis. Wie lange muss man hier sein, dass ich wählen kann? Wieso darf ich nicht bestimmen über die Regeln, die mein Leben beeinflussen? Ich kann nicht mitbestimmen. Das ist richtig scheiße. Ich bin nicht damit einverstanden, dass ich wirklich die deutsche Staatsbürgerschaft annehmen muss dafür. Wenn ich 20 Jahre hier leben, soll ich auch mitbestimmen können, finde ich. Und genauso blöd, […] ich kann in Kanada auch nicht wählen. […] Also ich bin auch dort ausgeschlossen. […] Und das heißt, ich darf nicht mitbestimmen in den zwei Orten, die mir am meisten bedeuten, was für Politik passiert. […] Politik beeinflusst alle Lebensbereiche. Und ich finde das extrem wichtig, dass man da mitreden kann. […] Ich bin Kanadierin, aber irgendwie auf der wichtigen Ebene der Politik, irgendwie nicht. Und ich bin Deutsche, aber […] dann nicht wichtig genug, um hier mitzubestimmen. [...] Ich bin wirklich sauer darüber, dass ich nirgends wählen kann" (IP4, #00:57:29-#01:01:51).

The interviews indicate that there are different strategies and motivations to actively engage and participate. Engagement and participation not only embed an individual in a local context on a social and political level, but also contribute significantly to a sense of being acknowledged and belonging, which are crucial for the sense of *Heimat*. Thus, for many people, *Heimat* depends on the possibility to actively contribute. However, legal framework conditions, such as the right to vote, which is tied to citizenship, can limit these possibilities.

4.3.2 Trans- and Plurilocal Practices

The process of creating a familiar place and establishing a *Heimat* can be accompanied by a process of disembedding from the places of growing up due to spatial and temporal distance from them. Depending on the experiences and intensity of

emotional attachment to places and the motivation to migrate, individuals dis-
embed from places and embed themselves in other places to a greater or lesser
extent. However, a strong disembedding can only be observed in the interview
with IP3, who intends to leave his past behind due to his negative experiences,
does not feel any (positive) emotional attachment, and has cut off most social
contacts in order to fully focus on his new life. Disembedding in the sense that
individuals focus mainly on the place they currently live in can also be found in
some other interviews (IP6, IP7, IP8, and IP9). For these participants, the connection to the places in their country of origin is limited to the closest social
contacts, such as family members, relatives, and friends, which they strive to
maintain. It is also evident in the other interviews that social relationships are
one of the main reasons for maintaining a connection to a place, through which
both embeddedness and an emotional connection are maintained. This translocal
embeddedness results from activities such as communication across nation-state
borders, which is why participants rely on modern means of communication. In
most cases, communication takes place via social media platforms and messen-
ger services, which is why an Internet connection and terminal devices such as
a smartphone or laptop are essential. IP6, for example, who has lived in different places in different countries and links her sense of *Heimat* mainly to people,
describes her laptop and smartphone as important reference points in her life, as
these devices represent a connecting point to people who are important to her.
Something similar can be seen in the interviews with other participants (IP1,
IP2, IP4, IP7, and IP12) who indicate more or less regular communication with
friends and family members. On the one hand, communication serves to get an
insight into their lives and to get 'up to date', and on the other hand, topics can
be discussed as well as practices exercised (speaking the native language, discussions, etc.), which find no or only little space in their everyday lives. However,
communication is impeded, for example, by large spatial distances, which involve
a time shift. Also by the lack of physical presence and thus limited possibilities
of joint activities, social relations are often limited only to the closest friends and
family members. However, translocal embeddedness through the maintenance of
social connections is not only expressed through communicative practices, but
also through the transfer of resources. This is evident, for example, in the interview with IP7, who is able to financially support her family in Colombia through
her well-paid job in Germany, which is very important to her. Also IP8, who
moved from Portugal to Austria, states to support her mother, who is her main
connection to Portugal.

Media and politics

In addition to maintaining social relations across borders, other translocal practices emerge from the interviews that maintain the connections to places in other countries and simultaneously embed the participants in different local contexts. This mainly includes media consumption (IP1, IP2, IP4, IP5, and IP11) and (aspired) political participation (IP4, IP5, IP10, and IP11). In terms of translocal media consumption, participants report that they use national and international news coverage to inform themselves about current events in their country of birth or in specific regions and cities. Apart from that, the participants state that they follow certain social media channels, listen to certain radio stations and music or watch certain TV series. On the one hand, participants consume these media contents because they are closer to their personal preferences than what is offered in the place where they currently live, and thus their connection to their country of origin finds expression in their everyday lives. And on the other hand, this maintains a connection by following and consuming current popular media products that can be relevant in exchanges with other people. In some interviews it becomes clear how important it is to be able to talk about seemingly banal everyday topics such as TV shows, movies or music in exchange with others (IP1 and IP4), as these can express and create a sense of belonging. From Germany via social media, IP1 therefore tries to follow what her friends and other people of the same age in Greece are currently talking about and what is 'in' at the moment, in order to always be up to date and thus to be able to keep up with conversations.

While some participants only occasionally follow the political events via news coverage in their country of origin or have no interest in doing so or do not aim to participate in political matters, because they have moved their center of life (IP7 and IP8) or consider their influence to be too small (IP3 and IP9), other participants state that they would like to participate politically (IP4) or already do so (IP5, IP10, and IP11). As mentioned above, IP4 is interested in political issues in Germany, where she currently lives, and in Canada, where she grew up, and therefore would like to be able to actively participate in politics by voting, but is not permitted to vote in either country due to legal regulations. IP5, on the other hand, has not only the German but also the Slovenian and Croatian citizenship, and thus also the right to vote in each country, which she also exercises. Due to her negative migration experiences, but also with regard to the Yugoslavian war from which she had to flee with her parents, she has developed a critical awareness in relation to nationalism and other inhuman tendencies, which is why she not only critically follows the political events in each country, but is also politically engaged, for example in relation to the situation of refugees in the Balkan region. The situation is different for the couple IP10 and IP11. Both neither own the German citizenship, nor expressed in the interview any desire to obtain it. Their Lithuanian citizenship allows them to vote in their country of origin, which is also very important to both of them. Both say that they follow political events and that it is important for them to participate, even though they no longer live there. IP11 in particular feels a strong urge to advocate for 'his country', which is why he is a member of a Lithuanian association as well as a Lithuanian investors club. He has noticed that since he and other 'like-minded people' no longer live in Lithuania, they are much more interested and involved in the political events in the country. The distance has changed his view of the country in a positive way. He also states that it is a big dream of his to be financially independent and to concentrate completely on his economic and political engagement for Lithuania.

Growing up plurilocal

For most participants, there was a significant shift in the center of life from one country to another. The extent of translocal practices and the effort to maintain connections to the places vary among the participants. Thus, cross-border plurilocal embeddedness also varies among participants. However, for the sisters IP1 and IP2, a determination of the center of life is not as clear as for other participants, as there has been more than one migration between two countries in their lives, and through their family since birth they have lived a plurilocal lifestyle encompassing two countries. While most participants can clearly designate which country they come from and where they currently live, IP1 and IP2 cannot clearly determine this difference. Both grew up bilingual and regularly flew to Greece in the summers when they lived in Germany, and flew to Germany when they lived in Greece. They still do that today. Both maintain contact with family members and friends in both countries and consume German as well as Greek media. IP1 attended school in both the German and Greek school systems, and also attended a German school in Greece in order to study in Germany afterwards. IP2, on the other hand, only attended schools in the Greek school system, although she also attended a Greek school after her second move to Germany in order to study in Greece afterwards. Such biographies, which demonstrate a plurilocal embedding of individuals already at an early age, cannot be subsumed under the term translocality, since the activities of these individuals do not emanate from one distinct center of life, but from at least two. Thus, while most of the participants operate mainly in the country they moved to and partly maintain their embeddedness in the country they left to some extent through translocal practices, IP1 and IP2 are exceptions who practice a plurilocal life.

4.4   (Re)negotiations of the Individual Sense of *Heimat*

The participants' biographies and especially their migration experiences are very different, and so is their personal sense of *Heimat*. The development of a sense of *Heimat* is an ongoing dynamic process that is shaped by a multiplicity of experiences. In this process, the factor of time plays a very significant role, since the age, time of migration, length of stay and future perspectives but also the retrospective have an influence on the negotiation as well as renegotiation of the personal sense of *Heimat*. The interviews show that the participants have developed different conceptions and senses of *Heimat* in their lives so far. These can be divided into two categories—which in some cases also overlap. In the following, the feeling of being in-between will be discussed, which can be seen as the result of (re)negotiations in relation to the sense of belonging and self-location (section 4.4.1). And on the other hand, it will be shown that *Heimat* can also be understood as an aspired ideal condition, which is not necessarily tied to a specific place, but the possibility in places to freely express the self (section 4.4.2).

4.4.1   Being in the In-Between

The process of *Heimat*-making involves a variety of practices and experiences that confront the individual with the unfamiliar and are accompanied over time by processes of familiarization, defamiliarization, as well as maintenance. Through temporal as well as spatial distance, perspectives on places change so that the relationship between 'here and there' and 'old and new' is renegotiated and reevaluated. This mainly concerns the self-location of an individual, meaning questions of identity and belonging, as can be seen in the interviews. For the participants who have not lived in the country where they grew up for several years, there were often certain key events that initiated a process of reflection on their own identity and belonging—sometimes accompanied by an (unpleasant) feeling of inner conflict. Some participants report, for example, that when they revisited their country of origin and the places they grew up for the first time after their long-term relocation, they realized how much they had become familiar with the place where they live now and that they had developed a sense of *Heimat* to a certain degree (IP4, IP6, IP7, IP10, IP11, and IP12). Although they still feel a connection to the places of their childhood and adolescence and the people there, and occasionally feel a nostalgic longing, the participants note that their center of life has shifted, the places where they grew up have changed, and that there are other aspects that they dislike. For these reasons, moving back tends to be out of the question. For example, IP4 notes that she realized in Germany how much she was shaped by her socialization in Canada—"the fish only knows it's a fish when it's out of water" (IP4, #01:03:58-#01:07:01)—and therefore she identifies as a Canadian. However, although she feels a strong emotional attachment to the places of her childhood and adolescence and Canada is part of her sense of *Heimat*, she also realized on her first visit to Canada in a long time, that she does not necessarily want to live there anymore and it would be very difficult for her to leave Germany. IP10 and IP11 also both identify strongly with their country of birth and describe themselves as Lithuanian. However, they also indicate that it would be difficult for them to adjust to living in Lithuania again.

The interviews indicate that the process of *Heimat*-making, including efforts to maintain existing person-place bonds, is situated in a field of tension between familiarization/familiarity and defamiliarization/alienation. This simultaneity of familiarity and alienation is perceived very differently by the participants. For example, it can provoke an uncomfortable feeling of inner conflict, as was the case for IP7 when she visited Colombia. She reports that she initially felt 'lost' and did not feel a clear sense of belonging anymore. She describes this sense of non-belonging as very uncomfortable, causing a process of self-reflection and renegotiation of her sense of *Heimat*:

"Also es war schon ein komisches Gefühl dahinzugehen und zu sagen, 'ok ich gehöre nicht mehr hier hin', und dann wieder zurückzukommen, und zu sagen, 'ich gehöre auch nicht hier hin'. Und das war ein unangenehmes Gefühl und deswegen habe ich mit Leuten darüber gesprochen und habe viel nachgedacht: 'Was bedeutet das für mich'? 'Und was möchte hier überhaupt erreichen'? 'Und was bedeutet für mich Kolumbien'?" (IP7, #00:21:08-#00:21:43). "Es ist so dieses Gefühl von Nicht-Zugehörigkeit, wenn man nirgendwo dazugehört. Und dann, was mache ich mit meiner Zukunft, wenn ich nicht da sein kann und nicht hier sein kann?" (IP7, #00:23:11-#00:23:36).

IP12 is in a similar process at the time of the interview. She describes Germany as her *Heimat*. But she also wants to establish a *Heimat* in New Zealand. She says that although she misses some things in Germany, she feels uncomfortable and confined there, which is why she would be reluctant to move back. In New Zealand, however, she has not yet been able to develop a sense of *Heimat* as she feels that she has to invest a lot more time and work. She experiences an inner conflict that reflects her negotiation of *Heimat*:

"Ich bin mir nicht sicher, ob ich da je ankomme. [...] Also es ist ja auch was, was ich mir selbst erlauben muss. [...] Also so denke ich über Heimat, dass ich mir das selbst irgendwie geben muss. Und ich hoffe, dass ich da schon hinkomme. Aber gleichzeitig fühle ich mich schon zerrissen irgendwie so im Sinne von: 'Kann ich nur eine Heimat haben'? Also, 'kann ich vielleicht auch zwei Heimaten haben'?" (IP12, #00:09:41-#00:11:11).

And later she says:

"Also ich würde da auch sagen, 100 Prozent Heimat kannst du nie wirklich haben. Also, weil in Deutschland fühle ich mich nicht 100 Prozent heimisch, wegen der Menschen hauptsächlich, glaube ich. Und hier fühle ich mich nicht 100 Prozent heimisch, weil ich es mir noch nicht "verdient" habe, oder so. [...] Also ich bezweifle, dass dieses 100 Prozent Heimat irgendwann kommt. Also ich kann mich dem annähern [...]" (IP12, #00:29:41-#00:30:17).

The feeling that there is no '100 percent *Heimat*' is also explicitly expressed by other participants (IP1 and IP3), but it can also be found implicitly in all other interviews. On the one hand, this expresses the fact that *Heimat* is a positively charged space that is associated with safety, security, freedom, warmth, etc. Many participants are aware that this is an ideal, which does not exist. On the other hand, the statement that there is no 100 percent *Heimat* expresses that there is not one place that is 100 percent *Heimat* for people who are plurilocally embedded and feel connected to places and people across borders. There is always 'something' missing in one place.

Locating the self nowhere and anywhere
The perception that there is no 100 percent *Heimat* is also reflected in the participants' sense of belonging and identity. This perception is subject to a dynamic process and can change over time and through different experiences. In the interviews, it appears that especially in the transition phase from a short-term to a long-term stay, i.e. when a sense of *Heimat* towards the new place of residence in another country has not yet been established, the sense of non-belonging can prevail (IP4, IP6, IP7, and IP12). The self does not appear to be clearly locatable, as a strong identification with the places in the country of origin as well as in the current place of residence is disrupted by the process of *Heimat*-making. In this process, the initial factors that orient individuals in their understanding of *Heimat* and shape the desire to maintain connections 'collide' with migration experiences as well as temporal and spatial distance which influence familiarization as well as defamiliarization. Within this process, *Heimat* is constantly

renegotiated, which can engender an inner conflict in terms of the individuals' self-location. The importance of self-location for individuals is indicated by the fact that a lack of identification and the sense of non-belonging causes emotional stress, as indicated in the interviews with IP7 and IP12 quoted above. However, the feeling of belonging nowhere can also be exchanged with the feeling of belonging anywhere. This is particularly evident in the following statement from IP6:

> "Es ist lustig, weil irgendwie manchmal fühle ich mich, als ob ich überall dazuge-
> höre. Ich denke mir, da habe ich Freunde und da habe ich Freunde. Ich kann eigentlich
> überall hingehen und kann irgendwo Anschluss finden. Und auf der anderen Seite
> denke ich mir manchmal, ich gehöre überhaupt nirgendswo dazu" (IP6, #00:14:25-
> #00:15:50).

Other participants also do not clearly locate themselves and state that they could live anywhere (IP3, IP8, IP9, and IP11). On the one hand, this is related to their motivations to leave one country, and on the other hand, it is confirmed by their positive experiences of creating their new livelihood. For example, IP11, who although he feels a strong attachment to his country of origin, calls it his *Heimat* and is politically committed to it, describes himself as a "world citizen" (IP11, #00:30:53-#00:31:05). He refers to Germany as his 'chosen *Heimat*' [*Wahlheimat*], which can be established in other places as well. However, it is clear from all the interviews that the self cannot be located anywhere, and thus the world as *Heimat* remains an ideal. An individual can only develop a sense of *Heimat* to a limited number of places, as this requires an embedding through social relationships, routinized practices, (positive) experiences as well as an investment of commitment and time. In addition, not every place in the world is 'a place to be', as all participants confirm that certain factors, such as preferred climatic conditions, the possibility of freely expressing oneself, human rights, etc., are crucial. Thus, an 'omnilocal' self-location should be viewed critically and rather be understood as a wishful thinking. However, a sense of *Heimat* and thus the emotional attachment to places and the sense of belonging can be plurilocal, which is confirmed in most of the interviews.

Locating the self plurilocal

Most participants relate their sense of *Heimat* through their everyday practices as well as their emotional attachment to at least two different places in two different countries. This is also often reflected in their sense of belonging and identity. Thus, they are embedded in multiple local contexts simultaneously. It is noticeable, on the one hand, that this sense of *Heimat* is subject to a dynamic process, as already mentioned, and is thus constantly renegotiated. And on the other hand, simultaneity does not necessarily imply a balance between the places of *Heimat*. This means, for example, that some participants make a distinction between *Heimat* as the place of origin and the new *Heimat* they have chosen for themselves. A distinction is also often made further at the national and local level. Thus, on the one hand, participants refer to concrete places and local particularities, but also to their national identity and their connection to a nation-state. Depending on the (previous) length of stay, reasons for the move and experiences, the emotional attachment and sense of belonging to the particular places and countries differ. As a result, certain tendencies can be discerned in terms of their embeddedness, but also in their understanding of their identity. For example, IP2 has lived in both Greece and Germany, has family in both countries, and calls both countries her *Heimat*. This is also reflected in her identity—she does not identify as one or the other, but as both. Nevertheless, she feels a stronger connection to Greece, which she explains by saying that she spent most of her childhood there, has more friends there, and can express herself better there in terms of language. Her sister IP1 feels similarly, although her biography is slightly different, as she is more embedded in both countries, experienced her childhood in both countries, has friends in both countries and can express herself very well in both languages. According to her, her experiences in Greece were more formative, so she feels a slightly stronger sense of belonging to Greece, although she also identifies as German and Greek. Other participants also show that they locate themselves in at least two countries, but often identify more strongly with one country (IP6, IP7, IP10, IP11, and IP12). IP5 is also plurilocally embedded but feels a stronger sense of belonging and an emotional connection to places in Croatia and Slovenia—she also identifies as Croatian, not as German. In her case, however, her plurilocal embeddedness is also expressed in a superordinate sense of *Heimat*. She also identifies as European and considers Europe to be her *Heimat*.

In the interview with IP4, in contrast, neither a clear tendency towards a country nor a lack of superordinate sense of belonging can be identified. She feels an emotional attachment to the places of her childhood as well as to Canada just as strongly as she does to Germany and the meaningful places in the country. She identifies as Canadian, but by living in Germany for several years and being embedded through social relations and everyday practices, she now also identifies as German. Her plurilocal sense of belonging and emotional attachment is most impressively reflected in her considerations on death and the question of the place of burial, which she raises at the very beginning of the interview. Initiated by the funeral of her father-in-law, she now also asks herself the question of where she would like to be buried later. This question is very difficult for her to answer:

> "Ich kann nicht sagen, wo ich lieber... Ich meine, gut, es gibt noch vieles in der Zukunft hoffentlich, was passieren wird zwischen jetzt und wenn ich sterbe. Aber wenn ich jetzt entscheiden müsste, möchte ich in Kanada noch hier bzw. ich möchte in beiden Orten untergebracht werden. Ich habe auch wirklich gesagt letzt-endlich, wenn es geht, würde ich am liebsten meine Asche aufteilen und in beide Orte. Aber das geht rechtlich nicht. Ich könnte nicht sagen, wo ich jetzt mehr hingehöre, weil ich in beiden Orten jetzt hingehöre. Nicht weil ich mich ortslos fühle, sondern weil ich irgendwie in beide gehöre und gerne in beiden wäre. Beide Orte sind mir eigentlich gleichgültig wichtig geworden jetzt" (IP4, #00:00:22-#00:05:17).

Later she repeats and emphasizes: "Für mich wäre das wirklich wichtig (!), dass ich in beiden Orten für die Ewigkeit liegen kann. Das ist wirklich sehr schwierig zu akzeptieren, dass das nicht erlaubt ist, weil das ist meine Identität. Beide sind meine Identität mittlerweile. Ich kann nicht ohne das andere Leben. Das ist wirklich, wer ich bin. Ich kann nicht ohne die andere Seite. Ich gehöre hierhin und ich gehöre genauso gut dorthin" (IP4, #00:47:46-#00:49:16).

The choice of a place of burial can be very revealing about an individual's sense of *Heimat*, as the burial in a certain place can be considered a 'homecoming' (c.f. GONEN, 2004) and thus expresses an individual's identity, belonging, and attachment to a place (c.f. AKKAYMAK and BELKHODJA, 2020; ZONTINI, 2015). This topic was not addressed in the other interviews. Firstly, because many participants focused on the past, present and near future in the interviews, but did not address their future life in high age. Only IP9 and IP12 mentioned that they expect to return to their country of origin after their working lives. And secondly, there was no key event so far, like IP4 experienced, which could initiate a process of thinking about death. The question of the place of burial, however, is probably not unambiguous for some of the other participants either.

4.4.2 *Heimat* as an Aspired Ideal Condition

All interviews show that *Heimat* carries positive connotations. In the *Heimat*, an individual feels safe and secure due to a familiar and acknowledging social environment. This feeling of safety and security is also enhanced by the opportunity to express the self freely. If this possibility or freedom is restricted by, for example, social or economic conditions, this can also affect the sense of *Heimat*. The interviews show that *Heimat* can be established in other places through active creation and investment of time. *Heimat* is not necessarily tied to a place, but rather means a desirable condition. Thus, *Heimat* is a dynamic concept determined by mobility and the search for an ideal condition. The desire for an ideal condition indicates that this concept of *Heimat* is a future-oriented one.

However, this condition cannot be considered in isolation from spatial aspects. The ideal condition of safety, security and freedom also depends on social, economic and legal conditions in the particular countries. In addition, the spatial is also reflected in the statements of the participants. For IP3, for example, freedom and mental health are necessary to be able to feel a sense of *Heimat*. Therefore, he says, "Heimat is where I am free and sane" (IP3, #00:03:06-#00:07:06). The 'where' is significant here because freedom and sanity are not given in all places for him. IP8 also links *Heimat* with the possibility of being able to express oneself freely:

"When you are in the place that you have space to be yourself, I think you can make any place feel home. I won't feel home in a place that I don't have space to be myself. So, if I have to move to a country where because of religion reasons or other reasons I cannot be myself, then I won't feel home for sure. If I am in a country where I would be discriminated about skin color, or about my style, or the way I speak, or if I am in a country that I cannot speak freely what I think, then I won't feel home" (IP8, #00:29:44-#00:30:58).

The aspiration to live in a place where self-expression is not restricted or threatened and all basic human needs are met is also present in the interviews with the other participants. This ideal place is called *Heimat*—all participants agree on this, although all also know that this ideal is difficult to achieve: "[...] there is nothing such as 100 percent Heimat" (IP3, #00:07:13-#00:07:33). However, the desire to find and create an ideal '*Heimat*-condition' is found especially among people who have not developed any or no strong emotional attachment to the places of their childhood and adolescence and the people living in these places, and thus have not been able to develop a sense of *Heimat* (IP3 and IP8). These participants also engage in little to no translocal practices and focus mainly on creating and maintaining their ideal condition. Nevertheless, their sense of *Heimat* is also shaped by the initial factors, which cannot be renounced in their seemingly ideal place, so that these have to be compensated for, or are missed. Individuals thus always 'carry' these formative initial factors with them, whereby *Heimat* is never only present- and future-oriented, but also simultaneously past-oriented.

Heimat: An Emotional Map of Biographical References

5

The study of migrant biographies proves to be very revealing about the phenomenon of *Heimat*. The experiences of international migrants provide insights into the factors that influence the (re)negotiation processes and thus also the development of the individual sense of *Heimat*. In the following, the results will be discussed with regard to the previously elaborated theoretical framework and research question. By drawing on the insights from the interviews and online diaries, as well as the geographical perspective on the phenomenon of *Heimat*, the spatial concept of *Heimat* will be reconceptualized and considered as an emotional map of biographical references. For this purpose, first the reflections of migrant biographies in the individual senses of *Heimat* will be discussed (section 5.1). Thereby it will be pointed out that migration generates different references, which converge in a sense of *Heimat*, but cannot be located in one place only. The second part will elaborate in how far international migration challenges static conceptions of *Heimat* (section 5.2). In consideration of simultaneities, the concept of *Heimat* as a biographical map will be evaluated on the meaning and interrelation of dynamism and stasis.

5.1   Reflections of Migrant Biographies in the Individual Sense of *Heimat*

Heimat is strongly related to the development of place attachments, as is evidenced by the initial factors of the familiar places that affect and orient the development of the sense of *Heimat* of individuals. The emergent initial factors of physical characteristics, familiar practices, social relationships, and formative experiences confirm that the 'vivid hauntings' (TUAN, 2012) of childhood and adolescence are constitutive of the sense of *Heimat*. These initial factors are

© The Author(s), under exclusive license to Springer Fachmedien Wiesbaden   89
GmbH, part of Springer Nature 2022
J. Andel, *Sense(s) of Heimat*, BestMasters,
https://doi.org/10.1007/978-3-658-38985-7_5

also reflected in the dimensions of place attachment according to SCANNELL and GIFFORD (2010). With regard to the sense of *Heimat* of the participants, it is clear that the factors and dimensions cannot be considered separately from each other. The physical characteristics of places, such as climatic conditions, landscape, ambient sounds, and so on, as well as familiar practices, for example, in terms of language, social interactions, or food, orient the individuals by serving as references for comparisons and preferences as these factors convey a sense of familiarity, security, and identification. For many people, these initial factors are an essential part of their sense of *Heimat*, which is why *Heimat* is often associated with the places of growing up (c.f. JOISTEN, 2003, p. 25). However, physical characteristics and familiar practices alone are not always sufficient to generate place attachment. More crucial for the development of a place attachment, however, are positive in-place-experiences, which are often associated with other people. *Heimat* is associated with positive emotions and thus presupposes a positive place attachment. This is reflected in the fact that people tend to not consider the places that are associated with negative experiences as their *Heimat*. While the physical characteristics as well as practices in these places orient the individual and can also be a part of the sense of *Heimat*, it is the social relations and experiences with other people that are formative and can either facilitate or inhibit the development of a sense of safety, security, familiarity, and belonging, which is required for a sense of *Heimat*. In particular, positive memories, such as the time and activities spent together with friends and family members at particular places, generate meaningful places and shape and orient the sense of *Heimat*. This is reflected, for example, in the 'proximity-maintaining behaviors' (SCANNELL and GIFFORD, 2010), which are those behaviors that express the desire to maintain the connection to places, and the people and practices which are connected to the places. Since *Heimat* is a concept with positive emotional connotations, individuals exclude singular negative experiences that they do not identify as part of their personal safe place. Thus, not all experiences from childhood and adolescence are part of *Heimat*. The exclusion of negative experiences and the imagination of *Heimat* are the result of cognitive processes initiated by spatial as well as temporal distancing (c.f. TUAN, 2016 [1974], p. 153). However, negative memories of places can equally influence person-place relationships and the sense of *Heimat*. In particular, when negative experiences predominate, such as the restriction of self-expression through a restrictive social environment or economic conditions, or emotionally stressful experiences of violence. As a consequence, the influence of negative experiences on the sense of *Heimat* is expressed in defining *Heimat* in terms of what is not *Heimat*. If a sense of safety, security and belonging cannot be developed, no positive emotional attachment to places can emerge. Consequently, this lack of positive identification is reflected in 'distance-seeking behaviors', such as moving away, cutting off social relations and avoiding a reconfrontation with what is associated with past experiences. Individuals, of course, always define *Heimat* in distinction to what is not *Heimat* (c.f. JOISTEN, 2003, p. 19). The difference is, however, that the non-*Heimat* on the one hand can stand in an opposite relation to positive experiences, which are part of the *Heimat*, and on the other hand can be based on negative experiences, which shape and orient the individual to draw consequences, to distance and to define, to create and to find *Heimat* contrary to the negative experiences.

The universality of *Heimat*
The initial factors of familiar places, such as physical characteristics, habitual and familiar practices, social relations, and particularly formative experiences, can be considered as universal influences in the childhood and adolescence of every person. The in-place-experiences in the places of growing up, as well as the subjective attribution of meaning, orient individuals in their personal definition of *Heimat*. As PROSHANSKY et al. write in regard to their concept of place identity, places provide a certain structure and influence the subjective sense of self (1983, p. 58). Places thus influence the development of self-identity and group-identity in the sense of belonging. Consequently, *Heimat* can be considered as a spatial manifestation of the sense of identity and belonging.

The significance of the initial factors (positive and negative experiences) as well as positive identification with places and the people within is also reflected in the three dimensions of the *Heimatgefühl* by MITZSCHERLICH (2019). One dimension of the sense of *Heimat* is the 'sense of community' (MITZSCHERLICH, 2019, p. 187), which is generated by positive identification in the sense of belonging. The sense of community requires the involvement of the individual, common practices as well as the acknowledgement of other people, as reflected in the interviews. A community includes not only family members and friends, but also other people with whom an individual identifies based on shared experiences and practices. MITZSCHERLICH describes another dimension of the sense of *Heimat* as a 'sense of control' (2019, p. 188). In this context, she considers control as behavioral security and the ability of action. The interviews reveal, that if the possibilities of free action and free self-expression are restricted, this negatively affects the identification with a place. As the third dimension of the sense of *Heimat*, MITZSCHERLICH identifies the 'sense of coherence' (2019, p. 188). She thus describes a feeling that is to be understood as the result of subjective sense-making and interpretation of various experiences. The sense of coherence involves the certainty of being in the right place and the feeling of 'having arrived'. This feeling was also expressed in various ways in the interviews—as present, absent, or aspiration.

By examining *Heimat* through the lens of migration it is apparent that these three dimensions of MITZSCHERLICH (2019) describe the sense of *Heimat* as twofold. On the one hand, the dimensions explain an actual condition, and on the other hand, the three dimensions provide a framework for understanding *Heimat* as a future-oriented ideal condition. Here, a distinction can be made between the general desire to have a *Heimat* and the active *Heimat*-making process in a new and unfamiliar environment. The desire for a *Heimat* in the sense of a place that offers safety, security, and the freedom for self-expression can be considered a universal basic human need (c.f. COSTADURA et al., 2019b, p. 19; GREVERUS, 1972). The relevance of the sense of *Heimat* is also reflected in its effects on the emotional wellbeing and (mental) health of many people (c.f. LENGEN, 2019; AL- ALI and KOSER, 2002b, p. 7). Thus, individuals orient their actions toward creating this ideal condition in a place or finding a place that offers this condition. An individual's biography plays a significant role in this orientation. The aspiration to find a *Heimat* is particularly evident among people who predominantly associate negative experiences with the places where they grew up and therefore behave in a distance-seeking mode, since they have not developed a positive place attachment and thus no sense of *Heimat*. Consequently, these people strive for a sense of community, control, and coherence in another place. The striving for experiencing a sense of *Heimat* also turns out to be universal, since all participants strive for it or have striven for and developed it in different ways with their actions. In order to develop a sense of *Heimat*—in other words, to establish a *Heimat*—in a new and unfamiliar environment international migrants have to actively create it by socio-emotionally appropriating space (c.f. BRUNS and MÜNDERLEIN, 2019, p. 105). This active process of space appropriation and *Heimat*-making involves various practices that embed the individual in an environment, and thereby create a familiar place that attains personal meaning through a sense of community, control, and coherence. However, the familiarization and embedding of an individual is at first significantly shaped by initial migration experiences, which can either facilitate or impede the development of a sense of *Heimat* to a considerable extent. Prior knowledge, for example, in the form of language skills, pre-existing social relationships, or knowledge of everyday practices, can have a positive impact on the initial migration experiences, as the individual feels a certain familiarity with a place. This familiarity also occurs when international migrants identify similarities, for example, in the physical characteristics of places or social practices. As familiar places with which an individual identifies serve as reference here, the individual identifies similarities between the self which is already embedded in a place and the new place (c.f. SCANNELL and GIFFORD,

2010, p. 3). Identification, existing similarities and the feeling of familiarity ease the individual's orientation in the new environment and provide a degree of security in action. A lack of prior knowledge and major differences, for example in terms of everyday practices, can negatively affect the sense of control. Uncertainty and the feeling of strangeness can have a very negative effect on the emotional well-being of an individual and can cause a longing for the familiar, for example in the form of *Heimweh*. Future prospects similarly affect the processes of familiarization and embeddedness. While an unsettled legal status and limited opportunities to act create a sense of uncertainty, a granted legal status, financial security through a job, or personally important and supportive people provide an anchor point and positive future prospects. Assured conditions for residence, the freedom and security for action, and future plans, for example, pursuing a specific career or starting a family, can be considered as both indicators and preconditions of a sense of control and coherence. *Heimat* is thus always dependent on the social, economic, political, and legal framework conditions.

The social encounters in a new and unfamiliar environment turn out to be particularly formative migration experiences. They particularly affect the dimension of the sense of community. Social encounters are perceived as negative by an individual if, for example, they emphasize the externally identified foreignness, are rejecting or even hostile. Such encounters can generate moments of irritation and a feeling of exclusion, and can also have a lasting negative impact on the sense of belonging. In this context, it becomes apparent that identification with places and the people in them depends significantly on the acceptance and acknowledgement of belonging by other people. The self-location in the sense of identification with places and the feeling of belonging is thus not only determined from the internal perspective of the individual, but mainly from an external perspective of other (groups of) people (c.f. GÖTZ, 2010, p. 206). As other people define belonging by locating 'the other' and 'the foreign' and thus marking a boundary between *Heimat* and non-*Heimat* (c.f. MEE and WRIGHT, 2009, p. 772), the exclusive potential of the sense of *Heimat* of the people who are already embedded in a place becomes apparent. In contrast, migrants also experience positive social encounters characterized, for example, by kindness, support and love. Such positive and inclusive experiences foster (social) embeddedness, a sense of belonging, and the development of an emotionally positive person-place bond. It also emerges from the interviews that in terms of belonging, the local is strongly linked to the national. Individuals relate *Heimat* both to specific places to which positive memories, social relations and daily life are linked, and to the nation-state as a whole, which is mainly associated with a national identity in the sense of common language and other (assumed) commonalities such as shared experiences with regard to education, socialization, practices and so on. The self-location of an individual thus depends on the development of the sense of *Heimat*. On the one hand, it depends on the place of growing up, which is influenced by the initial factors. And on the other hand, this is influenced by the experiences of migration and the process of *Heimat*-making in another place.

The senses of community, control and coherence cannot be considered separately, as they are mutually dependent, and only together do they constitute the sense of *Heimat*. In the interviews, however, it becomes apparent that these can be expressed to different extents. The expression of the sense of *Heimat* and the individual dimensions depend on time, experiences, and subjective sense-making and interpretation. The interviews show, for example, that a person may have lived in a country for a very long time, is familiar with social practices and is confident in his or her behavior (sense of control), and can also identify positively with certain places and (groups of) people (sense of community), but can still feel that he or she has not arrived due to early and strongly formative experiences of exclusion (sense of coherence). Other people may rapidly feel they have arrived in the right place (sense of coherence) because of their motivation to leave one place and seek better living conditions elsewhere, even though they do not yet feel a sense of belonging and are not socially involved (sense of community) and feel an uncertainty because they are still restricted in their freedom of action (sense of control). In each case, the development of the individual sense of *Heimat* and the personal conception of *Heimat* reflects the biography, migration experiences as well as the current time in life of an individual. All other different strategies of embedding and familiarization as well as experiences of the *Heimat*-making process also affect the development of the sense of *Heimat*. From the interviews, it is noticeable that the embedding of an individual in a local context depends, on the one hand, on routinized practices or socio-political engagement, which is interpreted as participating in and contributing to society from both internal and external perspectives. And on the other hand, the *Heimat*-making process involves the establishment of social relations, as these are crucial for the embedding and creation of a familiar place.

An emotional map called *Heimat*

The migrant process of socio-emotional appropriation of space does not necessarily include a process of disembedding and defamiliarization from the places of growing up. Although, due to the temporal and spatial distance, the *Heimat*-making process is always accompanied by a process of detachment, it can be observed that many international migrants, depending on their socio-emotional attachment, simultaneously strive to maintain a connection to familiar places and people. This is reflected in proximity-maintaining behaviors such as translocal practices, like media consumption, communication with friends and family members, or political participation. Through these translocal practices, individuals maintain their connection to meaningful places and locate themselves plurilocally. The *Heimat*-making process is thus also plurilocal, as the sense of *Heimat* must be actively maintained in a distant place across borders. Proximity-maintaining behaviors, however, can also be reflected in individuals creating a familiar place by compensating for the unfamiliar, integrating and seeking the familiar. Not all international migrants act translocally and strive to maintain a connection to the place of origin, but still many orient to the initial factors, such as the native language, food, or social practices that provide them with a sense of security and familiarity.

The spatial references of the sense of *Heimat* are constantly renegotiated by the multifaceted and partly opposing processes. As a result, a feeling of being in-between can emerge. However, the place of origin and the place of arrival, the here and there, the past, present and future cannot be considered separately from each other and converge in the individual sense of *Heimat*. *Heimat* represents a personal unique emotional geography of biographical-spatial references and is thus shaped by various personal experiences and (re)negotiations. From the interviews as well as online diaries, it becomes clear that through migration, *Heimat* is rarely clearly locatable in one place. The sense of *Heimat* is usually not limited to being attached to one place only, but many different places are charged with meaning over time. Consequently, no place represents '100 percent' of *Heimat*, as international migrants locate themselves plurilocally through various references, socio-emotional attachments and translocal practices. *Heimat* does not exist in one place and the places that are part of the individual sense of *Heimat* cannot be considered separately from each other, as they all contribute to the development of the spatial self. From this, it can be concluded that *Heimat* represents a spatial expression of a compound of different but interrelated senses. Due to the plurilocal biographical reference points this spatial expression can be described as an 'emotional map' (MITZSCHERLICH, 2019, p. 190), which most vividly illustrates the 'simultaneities of geography' (KATRAK, 1996, p. 129) inherent in the concept of *Heimat*, especially from the lens of international migration. Emotional maps appear primarily in the field of participatory planning, where they serve as a tool (PÁNEK, 2018, p. 18). The method of emotional mapping is used to visualize perceptions and emotional experiences of people in certain places. In contrast to 'mental maps', which reflect people's spatial environment from a very subjective point of view and selective memories, emotional maps are based on a spatially accurate base map. This base map serves as a background for people's drawings of their emotional experiences and perceptions of a spatial environment (PÁNEK, 2018, p. 19). In this way, for example, emotions such as fear, insecurity, relaxation, or enjoyment can be mapped. However, *Heimat* does not represent an emotional map in this classical sense. The notion of *Heimat* rather presents the result of spatial and environmental cognition, like a mental map. This means that the individual knows, reflects and reconstructs the space in thought and attaches emotional meaning and values based on experiences, feelings, memories and social relations (c.f. KITCHIN, 1994, pp. 1f.). This process of mental mapping or, in this case, emotional mapping links all the meaningful places, cognitions and expressions of the self to each other and contributes to the spatialization of a sense, which finds expression in *Heimat* as a designation. *Heimat* is mapped through biography and is constantly re-mapped through renegotiation processes. This map simultaneously represents every reference point of the self-location of an individual on various scales and can thus be understood as an expression of the '100 percent'. However, *Heimat* should rather be understood as an imagined emotional map which is located in the inner self and is difficult to visualize in its entirety. Nevertheless, this does not preclude that a momentary sense of *Heimat* can be visualized through emotional mapping of individually significant spatial reference points.

5.2   Towards a Simultaneously Static and Dynamic Conception of *Heimat*

The experiences of international migrants suggest that the concept of *Heimat* represents an anthropological constant across cultures (c.f. RUNIA, 2020, p. 169). In a globalized world, characterized by changes in the relation between space and time as well as increased flows of people, information, and commodities (GIDDENS, 1991b, p. 64), the concept of *Heimat* is subject to similar changes. The emotional map of *Heimat* is constantly changing along the biography of individuals, which migrate across borders. The development of a sense of *Heimat* is subject to dynamic processes of constant renegotiation. Through processes of socio-emotional appropriation of space, new spatial references emerge. Simultaneously, it is apparent that, on the one hand, the physical presence of an individual in a place is important, as processes of defamiliarization and disembedding occur due to spatial and temporal distance, but, on the other hand, physical presence is transcended through translocal practices, and individuals thus maintain their embeddedness and locate themselves plurilocally. Through the simultaneity of spatial as well as temporal references and opposing processes, individuals constantly renegotiate and reevaluate *Heimat*, so that the sense of *Heimat* can decline, be created, be relocated, and be expressed from a distance in practices. *Heimat* is thus always dynamic.

However, the results from the interviews and online diaries indicate that *Heimat* is not only characterized by dynamism and that it is too simplistic to conceptualize *Heimat* as entirely dynamic. Stasis is equally important for the concept of *Heimat* and the development of a sense of *Heimat*. The interviews and diaries show that individuals often refer to certain (spatial) reference points in their sense of *Heimat*, which are fixed. This can be, for example, the place of birth or the parental home, or certain ritualized practices that do not change. Furthermore, the importance of stasis is also expressed in the factor of time. In order to appropriate space socio-emotionally, individuals usually have to 'invest' time and have to reside in that space for a certain period of time. A feeling of having arrived and sedentariness is both a condition and an expression of the sense of *Heimat*. This is also indicated by the sense of coherence as an essential dimension of the sense of *Heimat* (c.f. MITZSCHERLICH, 2019, p. 188). The socio-emotional appropriation of space is also reflected in the sense of belonging. In the places of growing up, this appropriation usually unfolds in an unconscious and unreflective way over a long period of time (TUAN, 1980, p. 8). TUAN refers here to rootedness (1980, p. 8)—a very static concept. In contrast, the appropriation of new and foreign spaces is subject to active engagement and conscious actions (TUAN, 1980, p. 8). Regardless of whether the *Heimat*-making process is unconsciously or consciously shaped and experienced by an individual, the development of a person-place bond and sense of *Heimat* requires a sense of security, stability, and coherence, as well as a certain retention time. The development of a sense of *Heimat* is thus characterized by an interplay of dynamism and stasis. To put it in terms of refigurative spaces according to LÖW et al. (2021a, 2021b): Through processes of globalization and especially through increased migration flows, it becomes visible that *Heimat* is situated in a field of tension between dynamism/change and stasis/stability. As a spatial expression of a conglomerate of experiences, *Heimat* is deeply marked by simultaneities. The individual emotional map of *Heimat* is simultaneously an expression of dynamic processes as well as static references. This is also reflected in the location of the self, which is seldom unambiguous but rather characterized by multiple references and simultaneities. Consequently, international migration challenges static conceptions of *Heimat*, but does not negate the importance of stasis in the development of the sense of *Heimat*.

Conclusions on *Heimat* in the Context of International Migration

6

This thesis started by highlighting the vagueness of the concept of *Heimat* due to its multiplicity. This multiplicity can be perceived as a result of the (German) history of the concept of *Heimat*, which is characterized by shifts in meaning as well as by the uses, appropriations, and intentions of different actors. Although *Heimat* is a multifaceted, time- and context-dependent, and highly subjective concept, certain contents of the concept recur repeatedly, so that the phenomenon can be circumscribed. The history of the concept as well as interdisciplinary research shows that it is particularly associated with the places of growing up, landscape images, and senses of belonging and identity. These connotations depict a rather static and exclusive understanding of *Heimat*, which indicates an ideological charging of the concept as the history shows. In order to demystify *Heimat*, to challenge a static monolithic understanding, and to underscore the relevance of the concept for geographic research, the phenomenon was approached through the lens of international migration.

For this purpose, first a theoretical framing from a geographical perspective was designed, which in concrete terms means that *Heimat* in this thesis was understood and discussed as a spatial concept. *Heimat* is the product of multidimensional processes, but in particular subjective experiences such as perceptions, emotions, feelings, memories, imaginations and other cognitive processes as well as practices of self-location play a major role in the construction of *Heimat*. As *Heimat* is often defined in relation to a sense of *Heimat*, this thesis approached the concept from the perspective of emotional geographies. A sense of *Heimat* presupposes a positive person-place bond. This is a result of place-making processes in which individuals appropriate space socio-emotionally by attaching meaning to it. The formation of place attachment depends on the dimension of the actors, psychological processes, and the physical place as the object of the attachment. In relation to *Heimat*, place attachment is expressed on an emotional level through

J. Andel, *Sense(s) of Heimat*, BestMasters, https://doi.org/10.1007/978-3-658-38985-7_6

positive feelings such as safety, security, and familiarity, but also in multi-layered and interrelated dimensions of feelings such as senses of community, control, and coherence. The sense of *Heimat* is always also an expression of identification with places and their characteristics, especially with the people inhabiting them. Thus, the characteristics of places also permeate the self, shaping identity and the sense of belonging. Since *Heimat* is not only a subjective feeling, but also requires active creation and maintenance through everyday, routinized, and habitual practices, *Heimat* can also be conceptualized as a practice of self-location. Practices embed individuals in a local context and express belonging in a performative dimension. Thus, *Heimat* is an expression of the spatial self.

In order to view the concept of *Heimat* through the lens of migration, *Heimat* as a socio-spatial reference point for individuals was placed in the context of globalization. Due to the altered relations of space and time as well as the increase of flows, especially of people, *Heimat* is subject to spatial refigurations. This means that *Heimat*, which is traditionally understood as static, is situated in a field of tension between stability and change. This field of tension is characterized by simultaneities, which implies that opposing processes are co-present and shape space simultaneously. The fact that *Heimat* is characterized by simultaneities is particularly visible in the context of migration. Migration is often juxtaposed with sedentariness, although the realities of life for many international migrants show that the relationship between the here and there is not unambiguous. The problematic nature of this dichotomization has been highlighted by the concept of transnational migration or translocality. Through cross-border translocal practices, international migrants embed themselves plurilocally and maintain their socio-emotional connections to places despite spatial distances and physical absence. Simultaneously, international migrants appropriate new spaces through *Heimat*-making processes, thereby constantly renegotiating the individual sense of *Heimat*.

Through qualitative research on migrant biographies and experiences that was conducted via interviews and online diaries, the reflections of these biographies and experiences in the sense of *Heimat* of international migrants could be elaborated. On the one hand, it emerges that *Heimat* in the sense of an emotional geography and practice of self-location can be considered as universal, since a sense of *Heimat* is important for the emotional well-being as well as the self-definition of many people. Thus the sense of *Heimat* is aspired to through various strategies in a new environment and attempted to be maintained in the places that have been left through translocal practices. On the other hand, it becomes evident that *Heimat* is not one place or several separate places, but it rather should be considered as an emotional map. This emotional map is the product of constant renegotiation processes along the individual biography. It reflects the various spatial references, such as emotional attachments, social relations, experiences, and habitual practices that are linked to different places. Consequently, *Heimat* or the sense of *Heimat* is subject to various interrelated and sometimes contradictory processes, such as de/familiarization and dis/embedding, and can change over time. *Heimat* is thus dynamic. Simultaneously, however, *Heimat* is also static, since certain (spatial) reference points in the emotional map do not lose their meaning for individuals and represent an essential part of the individual sense of *Heimat*. Furthermore, the investment of time in a place and a feeling of settledness are essential for the development of a sense of *Heimat*. The individual sense of *Heimat* of international migrants in the form of an emotional map is thus simultaneously a spatial expression of dynamic renegotiation processes, which are shaped by diverse experiences, as well as static references, which are often meaningful for the sense of *Heimat* despite temporal and spatial distance and processes of reevaluation.

Developing a more profound elaboration of the conceptualization of *Heimat* as an emotional map requires further research using a variety of methods. For instance, the method of emotional mapping could be used in order to encourage participants to actually visualize their sense of *Heimat*. Furthermore, it would be fruitful to compare the experiences and feelings related to *Heimat* of international migrants with those of national migrants as well as people who have had no experience with migration in their life in order to gain more insights into the various biographical factors that influence the development of the individual sense of *Heimat* as well as to understand the universal meaning of *Heimat* for people. In order to gain further insights into the relevance of the sense of *Heimat*, it is also worth examining its effect on people's (mental) health from a medical and psychological perspective. Even though much has already been researched and written on the topic of *Heimat*, it offers exciting questions and there is still a need for (interdisciplinary) research.

References

AKKAYMAK, G. and C. BELKHODJA (2020): "Does Place Matter? Burial Decisions of Muslims in Canada", *Studies in Religion/Sciences Religieuses*, 49(3), pp. 372–388.

AL-ALI, N. and K. KOSER (eds.) (2002a): *New Approaches to Migration? Transnational communities and the transformation of home* (Routledge Research in Transnationalism, vol. 3), London, Routledge.

AL-ALI, N. and K. KOSER (2002b): "Transnationalism, international migration and home", in: N. AL-ALI and K. KOSER (eds.), *New Approaches to Migration? Transnational communities and the transformation of home* (Routledge Research in Transnationalism, vol. 3), London, Routledge, pp. 1–14.

ANDERSON, B. (2006): *Imagined Communities: Reflections of the Origin and Spread of Nationalism*, London, Verso.

ANDERSON, K. and S.J. SMITH (2001): "Editorial: Emotional geographies", *Transactions of the Institute of British Geographers*, 26, pp. 7–10.

APPADURAI, A. (^{7}2005): *Modernity at Large: Cultural Dimensions of Globalization* (Public Worlds, vol. 1), Minneapolis, University of Minnesota Press.

APPADURAI, A. (1996): *Modernity at Large: Cultural Dimensions of Globalization* (Public Worlds, vol. 1), Minneapolis, University of Minnesota Press.

ARNOLD, G. (2016): "Place and space in home-making processes and the construction of identities in transnational migration", *Transnational Social Review*, 6(1–2), pp. 160–177.

ARMSTRONG, H. (2004): "Making the Unfamiliar Familiar: Research Journeys towards Understanding Migration and Place", *Landscape Research*, 29(3), pp. 237–260.

BASCH, L., N. GLICK SCHILLER, C. SZANTON BLANC (1994): *Nations Unbound: Transnational Projects, Postcolonial Predicaments, and Deterritorialized Nation-States*, Longhorn, Gordon and Breach.

BASTIAN, A. (1995): *Der Heimat-Begriff: Eine begriffsgeschichtliche Untersuchung in verschiedenen Funktionsbereichen der deutschen Sprache* (Reihe Germanistische Linguistik, vol. 159), Berlin, De Gruyter.

BAUMAN, Z. (2017): *Retrotopia*, Cambridge UK, Polity Press.

BAUSINGER, H. (1986): "Heimat in einer offenen Gesellschaft: Begriffsgeschichte als Problemgeschichte", in: J. KELTER (ed.), *Die Ohnmacht der Gefühle: Heimat zwischen Wunsch und Wirklichkeit*, Weingarten, Drumlin, pp. 89–115.

BAUSINGER, H. (2009): "Chamäleon Heimat – eine feste Beziehung im Wandel", *Schwäbische Heimat*, 60(4), p. 396.

BAUSINGER H. and K. KÖSTLIN (eds.) (1980): *Heimat und Identität: Probleme regionaler Kultur* (Studien zur Volkskunde und Kulturgeschichte Schleswig-Holsteins, vol. 7), Neumünster, Karl Wachholtz.

BECK, U. (2002): *Macht und Gegenmacht im globalen Zeitalter: Neue weltpolitische Ökonomie*, Frankfurt am Main, Suhrkamp Verlag.

BECK, U. and N. SZNAIDER (2006): "Unpacking Cosmopolitanism for the Social Sciences: S Research Agenda", *The British Journal of Sociology*, 57(1), pp. 1–23.

BELSCHNER, W., S. GRUBITZSCH, C. LESZCZYNSKI and S. MÜLLER-DOOHM (eds.) (1995): *Wem gehört die Heimat: Beiträge der politischen Psychologie zu einem umstrittenen Phänomen* (Politische Psychologie, vol. 1), Wiesbaden, Springer.

BHABHA, H.K. (1994): *The Location of Culture*, New York, Routledge.

BINDER, B. (2020): "Politiken der Heimat, Praktiken der Beheimatung, oder: Warum das Nachdenken über Heimat zwar ermattet, aber dennoch notwendig ist", in: D. BÖNISCH, J. RUNIA and H. ZEHSCHNETZLER (eds.), *Heimat Revisited: Kulturwissenschaftliche Perspektiven auf einen umstrittenen Begriff*, Berlin, De Gruyter, pp. 85–105.

BINDER, B. (2010): "Beheimatung statt Heimat: Translokale Perspektiven auf Räume der Zugehörigkeit", in: M. SEIFERT (ed.), *Zwischen Emotion und Kalkül: 'Heimat' als Argument im Prozess der Moderne* (Schriften zur Sächsischen Geschichte und Volkskunde, vol. 35), Leipzig, Leipziger Universitätsverlag, pp. 189–204.

BINDER, S. and J. TOŠIĆ (2005): "Refugees as a Particular Form of Transnational Migrations and Social Transformations: Socioanthropological and Gender Aspects", *Current Sociology*, 53(4), pp. 607–624.

BLICKLE, P. (2002): *Heimat: A Critical Theory of the German Idea of Homeland*, Rochester NY, Camden House.

BLUNT, A. (2007): "Cultural geographies of migration: mobility, transnationality and diaspora", *Progress in Human Geography*, 31(5), pp. 684–694.

BLUNT, A. (2005): "Cultural geography: cultural geographies of home", *Progress in Human Geography*, 29(4), pp. 505–515.

BLUNT, A. and R. DOWLING (2006): *Home* (Key Ideas in Geography), New York, Routledge.

BOA, E. and R. PALFREYMAN (2000): *Heimat: A German Dream; Regional Loyalties and National Identity in German Culture, 1890–1990*, Oxford, Oxford University Press.

BONDI, L. (2005): "Making connections and thinking through emotions: between geography and psychotherapy", *Transactions of the Institute of British Geographers*, 30(4), pp. 433–448.

BÖNISCH, D., J. RUNIA and H. ZEHSCHNETZLER (eds.) (2020a): *Heimat Revisited: Kulturwissenschaftliche Perspektiven auf einen umstrittenen Begriff*, Berlin, De Gruyter.

BÖNISCH, D., J. RUNIA and H. ZEHSCHNETZLER (2020b): "Einleitung: Revisiting 'Heimat'", in: D. BÖNISCH, J. RUNIA and H. ZEHSCHNETZLER (eds.), *Heimat Revisited: Kulturwissenschaftliche Perspektiven auf einen umstrittenen Begriff*, Berlin, De Gruyter, pp. 1–19.

BRINKMANN, F.T. and J. Hammann (eds.) (2019): *Heimatgedanken: Theologische und kulturwissenschaftliche Beiträge* (pop.religion: lebensstil – kultur – theologie, vol. 5), Wiesbaden, Springer VS.

BRUNS, D. and D. MÜNDERLEIN (2019): "Internationale Konzepte zur Erklärung von Mensch-Ort-Beziehungen", in: M. HÜLZ, M., O. KÜHNE and F. WEBER (eds.), *Heimat:*

Ein vielfältiges Konstrukt (RaumFragen: Stadt – Region – Landschaft, vol. 33), Wiesbaden, Springer VS, pp. 99–119.

BMI (2020): "Heimat & Integration: Integrationskurse", *Bundesministerium des Innern, für Bau und Heimat*, [website], https://www.bmi.bund.de/DE/themen/heimat-integration/int egration/integrationskurse/integrationskurse-node.html;jsessionid=22CF207809BBCAC 67469548205481 3D4.1_cid373, (accessed 25 August 2021).

BURNARD, P. (1991): "A method of analysing interview transcripts in qualitative research", *Nurse Education Today*, 11, pp. 461–466.

BVA (2021): "Doppelte Staatsbürgerschaft (Mehrstaatlichkeit) nach deutschem Recht", *Bundesverwaltungsamt*, [website], https://www.bva.bund.de/SharedDocs/Kurzmeldu ngen/DE/Buerger/Ausweis-Dokumente-Recht/Staatsangehoerigkeit/Sonstige_Meldun gen/DoppelteStaatsbuergerschaft.html, (accessed 31 August 2021).

CHU, W. (2012): "National Socialism and Hierarchical Regionalism: The German Minorities in Interwar Poland", in: C.-C.W. SZEJNMANN and M. UMBACH (eds.), *Heimat, Region, and Empire: Spatial Identities under National Socialism*, Basingstoke, Palgrave Macmillan, pp. 72–90.

CLARK, M. (2020): *Digital Diaries as Social Research Method*, [online video], https://www. youtube.com/watch?v=xRuxXp-ud54, (accessed 4 August 2021).

COHEN, D.J., L.C. LEVITON, N. ISAACSON, A.F. TALLIA and B.F. CRABTREE (2006): "Online Diaries for Qualitative Evaluation: Gaining Real-Time Insights", *American Journal of Evaluation*, 27(2), pp. 163–184.

CONRADSON, D. and D. MCKAY (2007): "Translocal Subjectivities: Mobility, Connection, Emotion", *Mobilities*, 2(2), pp. 167–174.

COSTADURA, E., K. RIES and C. WIESENFELDT (eds.) (2019a): *Heimat global: Modelle, Praxen und Medien der Heimatkonstruktion* (Edition Kulturwissenschaft, vol. 188), Bielefeld, transcript Verlag.

COSTADURA, E., K. RIES and C. WIESENFELDT (2019b): "Heimat global: Einleitung", in: E. COSTADURA, K. RIES and C. WIESENFELDT (eds.), *Heimat global: Modelle, Praxen und Medien der Heimatkonstruktion* (Edition Kulturwissenschaft, vol. 188), Bielefeld, transcript Verlag, pp. 11–42.

DE FINA, A. (2015): "Narratives and Identities", in: A. DE FINA and A. GEORGAKOPOULOU (eds.), *The Handbook of Narrative Analysis*, Hoboken, John Wiley & Sons, pp. 351–368.

DONIG, N., S. FLEGEL and S. SCHOLL-SCHNEIDER (eds.) (2009): *Heimat als Erfahrung, und Entwurf* (Gesellschaft und Kultur: Neue Bochumer Beiträge und Studien, vol. 7), Berlin, Lit Verlag.

DOUGLAS, M. (1991): "The Idea of a Home: A Kind of Space", *Social Research*, 58(1), pp. 287–307.

DUNCAN, J.S. and D. LAMBERT (2003): "Landscapes of home", in: J.S. DUNCAN, N.C. JOHNSON and R.H. SCHEIN (eds.), *A companion to cultural geography*, Oxford, Blackwell, pp. 382–403.

DUYVENDAK, J.W. (2011): *The Politics of Home: Belonging and Nostalgia in Western Europe and the United States*, Basingstoke, Palgrave Macmillan.

DÜRRSCHMIDT, J. (^{2}2004): *Globalisierung*, Bielefeld, transcript Verlag.

EASTHOPE, H. (2004): "A place called home", *Housing, Theory and Society*, 21(3), pp. 128–138.

EASTHOPE, H. (2009): "Fixed Identities in a Mobile World?: The Relationship between Mobility, Place and Identity", *Identities: Global Studies in Culture and Power*, 16(1), pp. 61–82.

ECKER, G. (2012): "Prozesse der „Beheimatung": Alltags- und Memorialobjekte", in: F. EIGLER and J. KUGELE (eds.), *„Heimat': At the Intersection of Memory and Space* (Media and Cultural Memory / Medien und kulturelle Erinnerung, vol. 14), Berlin, De Gruyter, pp. 208–225.

EGGER, S. (2020): "Mi Heimat es su Heimat: Beobachtungen zu einem Schlüsselthema der flüchtigen Moderne", in: D. BÖNISCH, J. RUNIA and H. ZEHSCHNETZLER (eds.), *Heimat Revisited: Kulturwissenschaftliche Perspektiven auf einen umstrittenen Begriff*, Berlin, De Gruyter, pp. 23–39.

EGGER, S. (2014): *Heimat: Wie wir unseren Sehnsuchtsort immer wieder neu erfinden*, München, Riemann.

EHRKAMP, P. and H. LEITNER (2006): "Guest editorial: Rethinking immigration and citizenship: new spaces of migrant transnationalism and belonging", *Environmental and Planning A*, 38, pp. 1591–1597.

EICHMANNS, G. and Y. FRANKE (eds.) (2013): *Heimat Goes Mobile: Hybrid Forms of Home in Literature and Film*, Newcastle upon Tyne, Cambridge Scholars.

EIGLER, F. (2012): "Critical Approaches to Heimat and the "Spatial Turn"", *New German Critique*, 39(1), pp. 27–48.

EIGLER F. and J. KUGELE (eds.) (2012a): *'Heimat': At the Intersection of Memory and Space* (Media and Cultural Memory / Medien und kulturelle Erinnerung, vol. 14), Berlin, De Gruyter.

EIGLER, F. and J. KUGELE (2012b): "Introduction: Heimat at the Intersection of Memory and Space", in: F. EIGLER and J. KUGELE (eds.), *'Heimat': At the Intersection of Memory and Space* (Media and Culture Memory / Medien und kulturelle Erinnerung, vol. 14), Berlin, De Gruyter, pp. 1–12.

EISENSTADT, S.N. (1999): "Multiple Modernities in an Age of Globalization", in: C. HONEGGER, S. HRADIL and F. TRAXLER (eds.), *Grenzenlose Gesellschaft?* (Verhandlungen des 29. Kongresses der Deutschen Gesellschaft für Soziologie, des 16. Kongresses der Österreichischen Gesellschaft für Soziologie, des 11. Kongresses der Schweizerischen Gesellschaft für Soziologie in Freiburg i.Br. 1998), Wiesbaden, VS Verlag für Sozialwissenschaften, pp. 37–50.

ELO, S. and H. KYNGÄS (2008): "The qualitative content analysis process", *Journal of Advanced Nursing*, 62(1), pp. 107–115.

ESCHER, A. (2006): "I am Jamaican, I have a Syrian passport, and I have a British passport, too", in: G. GLASZE and J. THIELMANN (eds.), *"Orient" versus "Okzident"? Zum Verhältnis von Kultur und Raum in einer globalisierten Welt* (Mainzer Kontaktstudium Geographie, vol. 10), Mainz, Geographisches Institut, Johannes Gutenberg-Universität, pp. 53–64.

EVANS, R. (2013): "Learning and knowing: Narratives, memory and biographical knowledge in interview interaction", *European journal of Research on the Education and Learning of Adults*, 4(1), pp. 17–31.

FAIST, T. (2012): "Transnational migration", *The Wiley-Blackwell Encyclopedia of Globalization*, https://onlinelibrary.wiley.com/doi/abs/doi.org/10.1002/9780470670590. wbeog910, (accessed 20 July 2021).

FAIST, T. (2010): "Towards Transnational Studies: World Theories, Transnationalism and Changing Institutions", *Journal of Ethnic and Migration Studies*, 36(10), pp. 1665–1687.

FILEP, C.V., M. THOMPSON-FAWCETT, S. FITZSIMONS, and S. TURNER (2015): "Reaching revelatory places: The role of solicited diaries in extending research on emotional geographies into the unfamiliar", *Area*, 47(4), pp. 459–465.

FRANKE, N.M. (2017): *Naturschutz – Landschaft – Heimat: Romantik als eine Grundlage des Naturschutzes in Deutschland*, Wiesbaden, Springer VS.

GEBHARD, G., O. GEISLER and S. SCHRÖTER (eds.) (2007): *Heimat. Konturen und Konjunkturen eines umstrittenen Konzepts*, Bielefeld, transcript Verlag.

GEORGIOU, M. (2006): *Diaspora, Identity and the Media: Diasporic Transnationalism and Mediated Spatialities*, Cresskill, Hampton Press.

GIDDENS, A. (1991a): *Modernity and Self-Identity: Self and Society in the Late Modern Age*, Cambridge, Polity Press.

GIDDENS, A. (1991b): *The Consequences of Modernity*, Cambridge, Polity Press.

GIERYN, T. (2000): "A space for place in sociology", *Annual Review of Sociology*, 26, pp. 463–496.

GILMARTIN, M. (2008): "Migration, Identity and Belonging", *Geography Compass*, 2(6), pp. 1837–1852.

GIRTLER, R. (⁴2001): *Methoden der Feldforschung*, Wien, Böhlau Verlag.

GIULIANI, M.V. (2003): "Theory of attachment and place attachment", in: M. BONNES, T. LEE and M. BONAIUTO (eds.), *Psychological theories for environmental issues*, Aldershot, Ashgate, pp. 137–170.

GLICK SCHILLER, N. (2010): "A global perspective on transnational migration: Theorising migration without methodological nationalism", in: R. BAUBÖCK and T. FAIST (eds.), *Diaspora and Transnationalism: Concepts, Theories and Methods*, Amsterdam, Amsterdam University Press, pp. 109–129.

GLICK SCHILLER, N. (2007): "Transnationality", D. NUGENT and J. VINCENT (eds.), *A Companion to the Anthropology of Politics*, Malden, Blackwell Publishing, pp. 448–467.

GLICK SCHILLER, N., L. BASCH and C. SZANTON BLANC (1995): "From Immigrant to Transmigrant: Theorizing Transnational Migration", *Anthropological Quarterly*, 68(1), pp. 48–63.

GLICK SCHILLER, N., L. BASCH and C. SZANTON BLANC (eds.) (1992): *Towards a Transnational Perspective on Migration: Race, Class, Ethnicity, and Nationalism Reconsidered*, New York, New York Academy of Science.

GONEN, A. (2004): "Homecoming at Burial", *Horizons in Geography*, 60/61, pp. 421–426.

GÖTZ, I. (2010): "Nationale und regionale Identität: Zur Bedeutung von territorialen Verortungen in der Zweiten Moderne", in: M. SEIFERT (ed.), *Zwischen Emotion und Kalkül: ‚Heimat' als Argument im Prozess der Moderne* (Schriften zur Sächsischen Geschichte und Volkskunde, vol. 35), Leipzig, Leipziger Universitätsverlag, pp. 203–218.

GREVERUS, I.-M. (1979): *Auf der Suche nach Heimat*, München, C. H. Beck.

GREVERUS, I.-M. (1972): *Der territoriale Mensch: Ein literaturanthropologischer Versuch zum Heimatphänomen*, Frankfurt am Main, Athenäum.

HALBMAYER, E. and J. SALAT (2011): "Das ero-epische Gespräch", *Qualitative Methoden der Kultur- und Sozialanthropologie*, [website], https://www.univie.ac.at/ksa/elearning/cp/qualitative/qualitative-42.html, (accessed 4 August 2021).

HASHEMNEZHAD, H., A.A. HEIDARI and P.M. HOSEINI (2013): ""'Sense of Place" and "Place Attachment"", *International Journal of Architecture and Urban Development*, 3(1), pp. 5–12.

HASSE, J. (ed.) (2018): *Das Eigene und das Fremde: Heimat in Zeiten der Mobilität* (Neue Phänomenologie, vol. 30), Freiburg, Karl Alber.

HÄGELE, U. (2021): "Photography, Heimat, Ideology", in: C. WEBSTER (ed.), *Photography in the Third Reich: Art, Physiognomy and Propaganda*, Cambridge, Open Book Publishers, pp. 131–170.

HÄNEL, D. (2020): "Heimat – Anmerkungen aus der kulturwissenschaftlichen Praxis", in: D. BÖNISCH, J. RUNIA and H. ZEHSCHNETZLER (eds.), *Heimat Revisited: Kulturwissenschaftliche Perspektiven auf einen umstrittenen Begriff*, Berlin, De Gruyter, pp. 69–83.

HERMAND, J. and J. STEAKLEY (eds.) (1996): *Heimat, Nation, Fatherland: The German Sense of Belonging* (German Life and Civilization, vol. 22), New York, Peter Lang Publishing.

HERRMANN, H.P. (1996): ""Fatherland": Patriotism and Nationalism in the Eighteenth Century", trans. M. Sundell, in: J. HERMAND and J. STEAKLEY (eds.), *Heimat, Nation, Fatherland: The German Sense of Belonging* (German Life and Civilization, vol. 22), New York, Peter Lang Publishing, pp. 1–24.

HIDALGO, M.C. and B. HERNÁNDEZ (2001): "Place attachment: conceptual and empirical questions", *Journal of Environmental Psychology*, 21(3), pp. 273–281.

HILL, M. (2014): "Postmigrantische Alltagspraxen von Jugendlichen", in: E. YILDIZ and M. HILL (eds.), *Nach der Migration: Postmigrantische Perspektiven jenseits der Parallelgesellschaft* (Kultur & Konflikt, vol. 6), Bielefeld, transcript Verlag, pp. 171–192.

HOPF, C. (2004): "Qualitative Interviews: An Overview", in: U. FLICK, E. VON KARDORFF and I. STEINKE (eds.), *A Companion to Qualitative Research*, London, SAGE Publications, pp. 203–208.

HUIZINGA, R.P. and B. VAN HOVEN (2018): "Everyday geographies of belonging: Syrian refugee experiences in the Northern Netherlands", *Geoforum*, 96, pp. 309–317.

HÜLZ, M., O. KÜHNE and F. WEBER (eds.) (2019): *Heimat: Ein vielfältiges Konstrukt* (Raum-Fragen: Stadt – Region – Landschaft, vol. 33), Wiesbaden, Springer VS.

INALHAN, G. and E. FINCH (2004): "Place attachment and sense of belonging", *Facilities*, 22(5/6), pp. 120–128.

JOISTEN, K. (2003): *Philosophie der Heimat – Heimat der Philosophie*, Berlin, Akademie Verlag.

JONKE, G. (1994 [1969]): *Geometric Regional Novel*, trans. J.W. Vazulik, Normal IL, Dalkey Archive Press.

KATRAK, K.H. (1996): "South Asian American Writers: Geography and Memory", *Amerasian Journal*, 22(3), pp. 121–138.

KAUPPI N. and M.R. MADSEN (eds.) (2013): *Tansnational Power Elites: The New Professionals of Governance, Law and Security*, London, Routledge.

KITCHIN, R.M. (1994): "Cognitive Maps: What are they and why study them?", *Journal of Environmental Psychology*, 14, pp. 1–19.

KLOSE, J. (ed.) (2013): *Heimatschichten: Anthropologische Grundlegung eines Weltverhältnisses*, Wiesbaden, Springer VS.

KOSER, K. (2007): *International Migration: A Very Short Introduction*, Oxford, Oxford University Press.

KÖSTLIN, K. (2010): "Heimat denken: Zeitschichten und Perspektiven", in: M. SEIFERT (ed.), *Zwischen Emotion und Kalkül: ‚Heimat' als Argument im Prozess der Moderne* (Schriften zur Sächsischen Geschichte und Volkskunde, vol. 35), Leipzig, Leipziger Universitätsverlag, pp. 21–38.

KÜCK, S. (2021): *Heimat und Migration: Ein transdisziplinärer Ansatz anhand biographischer Interviews mit geflüchteten Menschen in Deutschland* (Sozial- und Kulturgeographie, vol. 43), Bielefeld, transcript Verlag.

LEHNERT, K. and B. LEMBERGER (2013): "Die Un-Ordnung neu denken – Probleme der Kategorisierung von ‚Migration' und Fragen an eine zukünftige Migrationsforschung", in: M. KLÜCKMANN and F. SPARACIO (eds.), *Spektrum Migration: Perspektiven auf einen alltagskulturellen Forschungsgegenstand*, Tübingen, TVV, pp. 91–110.

LENGEN, C. (2019): "Heimat und mentale Gesundheit: Wie *place identity* unser Heimatgefühl und Wohlbefinden beeinglusst", in: M. HÜLZ, M., O. KÜHNE and F. WEBER (eds.), *Heimat: Ein vielfältiges Konstrukt* (RaumFragen: Stadt – Region – Landschaft, vol. 33), Wiesbaden, Springer VS, pp. 121–146.

LESZCZYNSKA-KOENEN, A. (2019): „Heimat ist kein Ort", in: R. HAUBL and H.-J. WIRTH (eds.), *Grenzerfahrungen: Migration, Flucht, Vertreibung und die deutschen Verhältnisse*, Gießen, Psychosozial-Verlag, pp. 159–180.

LEVITT, P. (2004): "Transnational Migrants: When "Home" Means More Than One Country", *The Online Journal of the Migration Policy Institute*, https://www.migrationpolicy. org/article/transnational-migrants-when-home-means-more-one-country, (accessed 28 July 2021).

LOW, S.M. (1992): "Symbolic ties that bind", in: I. ALTMAN and S.M. LOW (eds.), *Place attachment*, New York, Plenum Press, pp. 165–185.

LÖW, M. and H. KNOBLAUCH (2021): "Raumfiguren, Raumkulturen und die Refiguration von Räumen", in: M. LÖW, V. SAYMAN, J. SCHWERER and H. WOLF (eds.), *Am Ende der Globalisierung: Über die Refiguration von Räumen* (Re-Figuration von Räumen, vol. 1), Bielefeld, transcript Verlag, pp. 25–57.

LÖW M., V. SAYMAN, J. SCHWERER and H. WOLF (eds.) (2021a): *Am Ende der Globalisierung: Über die Refiguration von Räumen* (Re-Figuration von Räumen, vol. 1), Bielefeld, transcript Verlag.

LÖW, M., V. SAYMAN, J. SCHWERER and H. WOLF (2021b): "Am Ende der Globalisierung: Über die Refiguration von Räumen", in: M. LÖW, V. SAYMAN, J. SCHWERER and H. WOLF (eds.), *Am Ende der Globalisierung: Über die Refiguration von Räumen* (Re-Figuration von Räumen, vol. 1), Bielefeld, transcript Verlag, pp. 9–22.

MAROTZKI, W. (2004): "Qualitative Biographical Research", in: U. FLICK, E. VON KARDORFF and I. STEINKE (eds.), *A Companion to Qualitative Research*, London, SAGE Publications, pp. 101–107.

MASON, J. (2002): "Qualitative interviewing: Asking, listening and interpreting", in: T. MAY (ed.), *Qualitative Research in Action*, London, SAGE Publications, pp. 225–241.

MASSEY, D. (2005): *For Space*, London, SAGE Publications.

MASSEY, D. (1995): "The conceptualization of place", in: D. MASSEY and P. JESS (eds.), *A Place in the World?: Places, Cultures and Globalization*, Oxford, Oxford University Press, pp. 45–85.

MAYRING, P. ([12]2010): *Qualitative Inhaltsanalyse: Grundlagen und Techniken*, Weinheim, Beltz Verlag.

MEE, K. and S. WRIGHT (2009): "Guest editorial: Geographies of belonging", *Environmental and Planning A*, 41, pp. 772–779.

METH, P. (2003): "Entries and omissions: using solicited diaries in geographical research", *Area*, 35(2), pp. 195–205.

MITCHELL, D. (2000): *Cultural Geography: A Critical Introduction*, Malden, Blackwell Publishing.

MITCHELL, K. (2003): Cultural geographies of transnationality. In: K. ANDERSON, M. DOMOSH, S. PILE and N. THRIFT (eds.), *Handbook of Cultural Geography*, London: SAGE Publications, pp. 74–87.

MITZSCHERLICH, B. (2019): "Heimat als subjektive Konstruktion: Beheimatung als aktiver Prozess", in: E. COSTADURA, K. Ries and C. WIESENFELDT (eds.), *Heimat global: Modelle, Praxen und Medien der Heimatkonstruktion* (Edition Kulturwissenschaft, vol. 188), Bielefeld, transcript Verlag, pp. 183–195.

MITZSCHERLICH, B. (1997): *„Heimat ist etwas, was ich mache": Eine psychologische Untersuchung zum individuellen Prozess von Beheimatung*, Pfaffenweiler, Centaurus-Verlagsgesellschaft.

MOOSMANN, E. (ed.) (1980): *Heimat: Sehnsucht nach Identität*, Berlin, Ästhetik-und-Kommunikation-Verlags-GmbH.

MORLEY, D. (2001): "Belongings: Place, space and identity in a mediated world", *European Journal of Cultural Studies*, 4(4), pp. 425–448.

NOHL, A.-M. (2010): "Narrative Interview and Documentary Interpretation", in: R. BOHNSACK, N. PFAFF and W. WELLER (eds.), *Qualitative analysis and documentary method in international educational research*, Opladen, B. Budrich, pp. 195–217.

PÁNEK, J. (2018): "Emotional Maps: Participatory Crowdsourcing of Citizens' Perceptions of Their Urban Environment", *Cartographic Perspectives*, 91, pp. 17–29.

PAPASTERGIADIS, N. (1998): *Dialogues in the Diaspora: Essays and Conversations on Cultural Identity*, London, Rivers Oram Press.

PERRINO, S. (2015): "Chronotopes: Time and Space in Oral Narrative", in: A. DE FINA and A. GEORGAKOPOULOU (eds.), *The Handbook of Narrative Analysis*, Hoboken, John Wiley & Sons, pp. 140–159.

PILE, S. (2010): "Emotions and affect in recent human geography", *Transactions of the Institute of British Geographers*, 35(1), pp. 5–20.

PROSHANSKY, H.M. (1978): "The City and Self-Identity", *Environment and Behavior*, 10(2), pp. 147–169.

PROSHANSKY, H.M. and A.K. FABIAN (1987): "The Development of Place Identity in the Child", in: C.S. WEINSTEIN and T.G. DAVID (eds.), *Spaces for Children: The Built Environment and Child Development*, New York, Plenum, pp. 21–40.

PROSHANSKY, H.M., A.K. FABIAN and R. KAMINOFF (1983): "Place-Identity: Physical World Socialization of the Self", *Journal of Environmental Psychology*, 3, pp. 57–83.

RALPH, D. and L. STAEHELI (2011): "Home and Migration: Mobilities, Belongings and Identities", *Geography Compass*, 5(7), pp. 517–530.

RANDERIA, S. (1999): "Jenseits von Soziologie und soziokulturelle Anthropologie: zur Verortung der nichtwestlichen Welt in einer zukünftigen Sozialtheorie", *Soziale Welt*, 50(4), pp. 373–382.

RAPPORT, N. and A. DAWSON (1998a): "The topic and the book", in: N. RAPPORT and A. DAWSON (eds.), *Migrants of Identity: perceptions of home in a world of movement*, Oxford, Berg, pp. 3–17.

RAPPORT, N. and A. DAWSON (1998b): "Home and Movement: A Polemic", in: N. RAPPORT and A. DAWSON (eds.), *Migrants of Identity: perceptions of home in a world of movement*, Oxford, Berg, pp. 19–38.

RECKWITZ, A. (2003): "Grundelemente einer Theorie sozialer Praktiken: Eine sozialtheoretische Perspektive", *Zeitschrift für Soziologie*, 32(4), pp. 282–301.

RISHBETH, C. and M. POWELL (2013): "Place Attachment and Memory: Landscapes of Belonging as Experienced Post-migration", *Landscape Research*, 38(2), pp. 160–178.

ROBERTS, E. (2012): "Family Photographs: Memories, Narratives, Place", in: O. JONES and J. Garde-Hansen (eds.), *Geography and Memory: Explorations in Identity, Place and Becoming* (Palgrave Macmillan Memory Studies, vol. 20), Basingstoke, Palgrave Macmillan, pp. 91–108.

ROBERTSON, R. (1995): "Glocalization: Time-Space and Homogeneity-Heterogeneity", in: M. FEATHERSTONE, S. LASH and R. ROBERTSON (eds.), *Global Modernities*, London, SAGE Publications, pp. 25–44.

ROLLINS, W. (1996): "Heimat, Modernity, and Nation in the Early Heimatschutz Movement", in: J. HERMAND and J. STEAKLEY (eds.), *Heimat, Nation, Fatherland: The German Sense of Belonging* (German Life and Civilization, vol. 22), New York, Peter Lang Publishing, pp. 87–112.

ROSE, G. (1995): "Place and Identity: A Sense of Place", in: D. MASSEY and P. JESS (eds.), *A Place in the World?: Places, Cultures and Globalization*, Oxford, Oxford University Press, pp. 87–132.

RÖMHILD, R. (2014): "Jenseits ethnischer Grenzen: Für eine postmigrantische Kultur- und Gesellschaftsforschung", in: E. YILDIZ and M. HILL (eds.), *Nach der Migration: Postmigrantische Perspektiven jenseits der Parallelgesellschaft* (Kultur & Konflikt, vol. 6), Bielefeld, transcript Verlag, pp. 37–48.

RUNIA, J. (2020): "Mobile Verwurzelung: Hybride Heimatkonzeptionen in Randa Jarrars *A Map of Home*", in: D. BÖNISCH, J. RUNIA and H. ZEHSCHNETZLER (eds.), *Heimat Revisited: Kulturwissenschaftliche Perspektiven auf einen umstrittenen Begriff*, Berlin, De Gruyter, pp. 167–187.

RUTHERFORD, J. (1990): "A Place Called Home: Identity and the Cultural Politics of Difference", in: J. RUTHERFORD (ed.), *Identity: Community, Culture, Difference*, London, Lawrence and Wishart, pp. 9–27.

SALMONS, J.E. (2016): *Doing Qualitative Research Online*, London, SAGE Publications.

SANDLER, W. (2012): "'Here Too Lies Our *Lebensraum*': Colonial Space as German Space", in: C.-C.W. SZEJNMANN and M. UMBACH (eds.), *Heimat, Region, and Empire: Spatial Identities under National Socialism*, Basingstoke, Palgrave Macmillan, pp. 148–165.

SANDU, A. (2013): "Transnational Homemaking Practices: Identity, Belonging and Informal Learning", *Journal of Contemporary European Studies*, 21(4), pp. 496–512.

SAVAGE, M., G. BAGNALL and B. LONGHURST (2005): *Globalization and Belonging*, London, SAGE Publications.

SCANNELL, L. and R. GIFFORD (2010): "Defining place attachment: A triapartite organizing framework", *Journal of Environmental Psychology*, 30, pp. 1–10.

SCHARNOWSKI, S. (2019): *Heimat: Geschichte eines Missverständnisses*, Darmstadt, wbg.

SCHMIDT, C. (2004): "The Analysis of Semi-structured Interviews", in: U. FLICK, E. VON KARDORFF and I. STEINKE (eds.), *A Companion to Qualitative Research*, London, SAGE Publications, pp. 253–258.

SCHRAMM, H. and N. LIEBERS (2019): "Heimat – das ist ein Gefühl", *M&K Medien & Kommunikationswissenschaft*, 67(3), pp. 259–276.

SCHULZ-SCHAEFFER, I. (2010): "Praxis, handlungstheoretisch betrachtet", *Zeitschrift für Soziologie*, 39(4), pp. 319–336.

SCHURR, C. (2014): „Emotionen, Affekte und mehr-als-repräsentationale Geographien", *Geographische Zeitschrift*, 102(3), pp. 148–161.

SEIFERT, M. (ed.) (2010a): *Zwischen Emotion und Kalkül: ,Heimat' als Argument im Prozess der Moderne* (Schriften zur Sächsischen Geschichte und Volkskunde, vol. 35), Leipzig, Leipziger Universitätsverlag.

SEIFERT, M. (2010b): "Das Projekt ,Heimat' – Positionen und Perspektiven", in: M. SEIFERT (ed.), *Zwischen Emotion und Kalkül: ,Heimat' als Argument im Prozess der Moderne* (Schriften zur Sächsischen Geschichte und Volkskunde, vol. 35), Leipzig, Leipziger Universitätsverlag, pp. 9–22.

STAEHELI, L.A. and C.R. NAGEL (2006): "Topographies of home and citizenship: Arab-American activists in the United States", *Environment and Planning A*, 38, pp. 1599–1614.

SVAŠEK, M. (2010): "On the Move: Emotions and Human Mobility", *Journal of Ethnic and Migration Studies*, 36(6), pp. 865–880.

SVAŠEK, M. and M. DOMECKA (2012): "The Autobiographical Narrative Interview: A Potential Arena of Emotional Remembering, Performance and Reflection", in: J. SKINNER (ed.), *The Interview: An Ethnographic Approach*, London, Bloomsbury Academic, pp. 107–126.

SZEJNMANN, C.-C.W. (2012): "'A Sense of *Heimat* Opened Up during the War.' German Soldiers and *Heimat* Abroad", in: C.-C.W. SZEJNMANN and M. UMBACH (eds.), *Heimat, Region, and Empire: Spatial Identities under National Socialism*, Basingstoke, Palgrave Macmillan, pp. 112–147.

SZEJNMANN, C.-C.W. and M. UMBACH (eds.) (2012): *Heimat, Region, and Empire: Spatial Identities under National Socialism*, Basingstoke, Palgrave Macmillan.

THERBORN, G. (2003): "Entangled Modernities", *European Journal of Social Theory*, 6(3), pp. 293–305.

TOMANEY, J. (2014): "Region and place 2: Belonging", *Progress in Human Geography*, 39(4), pp. 507–516.

TROISI, J.D. and S. GABRIEL (2011): "Chicken Soup Really Is Good for the Soul: "Comfort Food" Fulfills the Need to Belong", *Psychological Science*, 22(6), pp. 747–753.

TUAN, Y.-F. (2016 [1974]): "Space and Place: Humanistic Perspective", in: A. ESCHER and S. PETERMANN (eds.), *Raum und Ort* (Basistexte Geographie, vol. 1), Stuttgart, Franz Steiner Verlag, pp. 133–166.

TUAN, Y.-F. (2012): "Epilogue: Home as Elsewhere", in: F. EIGLER and J. KUGELE (eds.), *'Heimat': At the Intersection of Memory and Space* (Media and Cultural Memory / Medien und kulturelle Erinnerung, vol. 14), Berlin, De Gruyter, pp. 226–239.

TUAN, Y.-F. (2004): "Home", in: S. HARRISON, S. PILE and N. THRIFT (eds.), *Patterned ground: the entanglements of nature and culture*, London, Reaktion Books, pp. 164–165.

TUAN, Y.-F. (1980): "Rootedness versus Sense of Place", *Landscape*, 24, pp. 3–8.

TUAN, Y.-F. (1977): *Space and Place: The Perspective of Experience*, Minneapolis MN, University of Minnesota Press.

TWIGGER-ROSS, C.L. and D.L. UZZELL (1996): "Place and identity processes", *Journal of Environmental Psychology*, 16(3), pp. 205–220.

United Nations (2020a): "International Migration 2020 Highlights", *United Nations Department of Economic and Social Affairs*, [website], https://www.un.org/development/desa/pd/sites/www.un.org.development.desa.pd/files/undesa_pd_2020_international_migration_highlights.pdf, (accessed 26. July 2021).

United Nations (2020b): "International Migration 2020 Highlights", *United Nations Department of Economic and Social Affairs*, [website], https://www.un.org/en/desa/international-migration-2020-highlights, (accessed 26. July 2021).

URRY, J. (2000): *Sociology beyond Societies: Mobilities for the twenty-first century*, London, Routledge.

VON MOLTKE, J. (2005): *No Place Like Home: Locations of Heimat in German Cinema*, Berkeley and Los Angeles CA, University of California Press.

WARDENGA, U. (2002): "Räume der Geographie und zu Raumbegriffen im Geographieunterricht", *Wissenschaftliche Nachrichten*, 120, pp. 47–52.

WEBER, F., O. KÜHNE and M. HÜLZ (2019): "Zur Aktualität von 'Heimat' als povalentem Konstrukt – eine Einführung", in: M. HÜLZ, M., O. KÜHNE and F. WEBER (eds.), *Heimat: Ein vielfältiges Konstrukt* (RaumFragen: Stadt – Region – Landschaft, vol. 33), Wiesbaden, Springer VS, pp. 3–23.

WICKHAM, C.J. (1999): *Constructing Heimat in Postwar Germany: Longing and Belonging* (Studies in German Thought and History, vol. 18), Lewiston NY, The Edwin Mellen Press.

WICKRAMASINGHE, A.A.I.N. and W. WIMALARATANA (2016): "International Migration and Migration Theories", *Social Affairs*, 1(5), pp. 13–32.

WIMMER, A. and N. GLICK SCHILLER (2002): "Methodological Nationalism and Beyond: Nation-State Building, Migration, and the Social Science", *Global Networks*, 2, pp. 301–334.

YILDIZ, E. (2014): "Postmigrantische Perspektiven: Aufbruch in eine neue Geschichtlichkeit", in: E. YILDIZ and M. HILL (eds.), *Nach der Migration: Postmigrantische Perspektiven jenseits der Parallelgesellschaft* (Kultur & Konflikt, vol. 6), Bielefeld, transcript Verlag, pp. 19–36.

YILDIZ, E. and M. HILL (eds.) (2014a): *Nach der Migration: Postmigrantische Perspektiven jenseits der Parallelgesellschaft* (Kultur & Konflikt, vol. 6), Bielefeld, transcript Verlag.

YILDIZ, E. and M. HILL (2014b): "Einleitung", in: E. YILDIZ and M. HILL (eds.), *Nach der Migration: Postmigrantische Perspektiven jenseits der Parallelgesellschaft* (Kultur & Konflikt, vol. 6), Bielefeld, transcript Verlag, pp. 9–16.

ZONTINI, E. (2015): "Growing old in a transitional social field: Belonging, mobility and identity among Italian migrants", *Ethnic and Racial Studies*, 38(2), pp. 326–341.

Printed in the United States
by Baker & Taylor Publisher Services